SCRIBE
Melbourne • London

Answers to the world's
WEIRDEST QUESTIONS,
most PERSISTENT RUMOURS, and
UNEXPLAINED PHENOMENA

Mitchell Moffit and Greg Brown

Illustrations by Greg Brown, Jessica Carroll, and Mitchell Moffit

Scribe Publications
18–20 Edward St, Brunswick, Victoria 3056, Australia
2 John St, Clerkenwell, London, WC1N 2ES, United Kingdom

First published by Scribe 2015, reprinted 2015

This edition published by arrangement with Simon and Schuster, Inc.

Interior design Brian Chojnowski
Printed in Italy by L.E.G.O. S.p.A. Lavis

9781925106589 (Australian hardback)
9781922247940 (UK hardback)
9781925113815 (e-book)

A CIP record for this title is available from the British Library.

scribepublications.com.au
scribepublications.co.uk

This book is dedicated to our parents for their consistent
support and inspiration. Thank you for always
nurturing our curiosity of the world around us. We love you.

Contents

Resolving Persistent Questions, Rumours, and Weird Phenomena

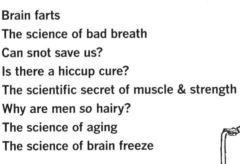

Body Talk

Hypotheticals

Sensory Perception

Hot Sex and Other Amorous Pursuits

Getting to the Bottom of Bad Behaviour

Dreaming, Waking, Napping, Sleeping

RESOLVING PERSISTENT QUESTIONS, RUMOURS, AND WEIRD PHENOMENA

DOES BEING COLD MAKE YOU SICK?

"Winter is coming," and we all remember our parents telling us to "put a jacket on to go outside or you'll catch a cold!" Of course, the common response is, "Don't be silly, being cold can't give you a cold!" Or can it? Who wins this battle?

There does seem to be a correlation between cold weather and sickness. Between 5 and 20 percent of Americans catch a cold or get the flu every year in late fall and winter. Not to mention, it's called a *cold*.

But there are some important things to consider. First and foremost: Colds and the flu are caused by viruses. If there aren't any around, you won't catch a cold—no matter how cold you get. It's as simple as that.

So why the correlation with frigid temperatures? Well, for one, people tend to stay indoors much more during the winter, which puts them in contact with more people. More people means more exposure opportunities for the pathogens to spread.

On top of this, humidity plays a role in the spread of some viruses. As the humidity decreases in the winter, not only does the virus spread more readily but the mucus in your nose dries out—mucus that would otherwise act as a protective barrier to pathogens.

Finally, vitamin D, which we get from the sun, is vital to our immune system function. The days are shorter in the winter and we're inside more, so we get much less vitamin D, which can have an adverse effect on health.

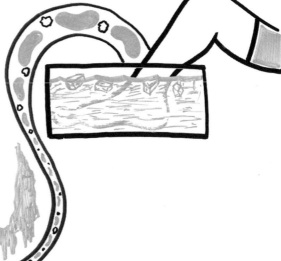

VITAMIN D

So you've proved your parents wrong . . . right? Not so fast!

While some studies have shown no correlation between getting sick and temperature, recent evidence suggests otherwise. One study put test subjects' feet into ice water and found that they were, indeed, more likely to develop common cold symptoms than those who didn't. The developing hypothesis behind these results is that cold temperatures cause blood vessel constriction, which slows the white blood cells from reaching the virus, ultimately inhibiting the immune response.

white blood cells

Levels of the hormone cortisol, which suppresses the immune system, also increase with temperature-induced stress. Furthermore, studies on both mouse and human airway cells found that immune reaction to the common cold virus is, in fact, temperature dependent. Warm infected cells are more likely to undergo programmed cell death, to limit the spread of infection.

Finally, studies of the virus itself have revealed a secret weapon of sorts. In winter temperatures, the virus's outer layer or envelope becomes much harder and acts like a shield. This allows it to spread from person to person much more easily.

But at warm temperatures, this layer is more of a gel, which is not quite tough enough to protect the virus against the elements. As a result, its spreading ability is compromised.

So maybe your parents weren't *so* wrong after all. Here's a happy compromise: Go outside more often, bundled up. You'll be more likely to get through the winter unscathed.

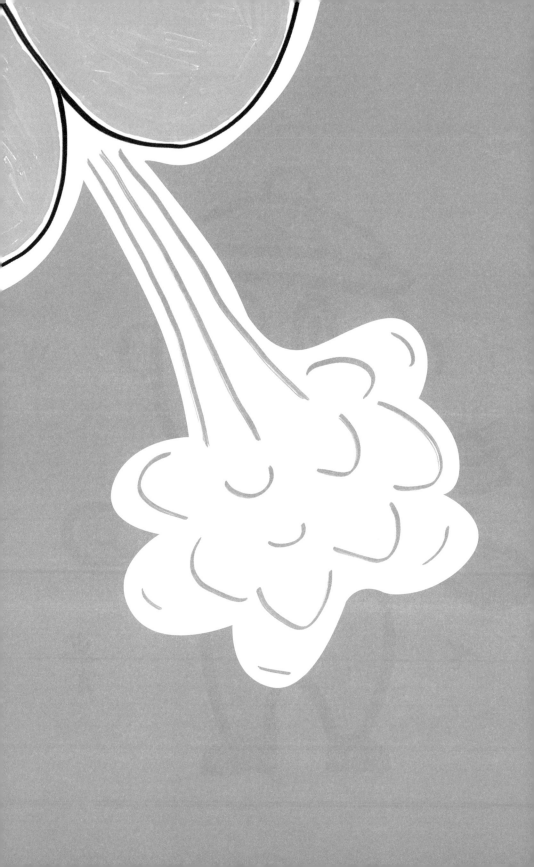

ARE SILENT FARTS MORE DEADLY?

The idea of "silent but deadly" farts may make you laugh, but is this faint flatulence really more potent than the boisterous wind you let loose? Are silent farts worse?

It may surprise you to learn that many farts are the result of swallowed air. Whether from chewing gum, drinking pop, or simply eating food, excess air enters your body and has to go somewhere.

While some is released through burping, the rest ends up in the digestive system and eventually comes out your other end. But when expelled, this *gas* is made of mostly nitrogen, hydrogen, and carbon dioxide—all of which are odorless. Which explains why some farts are obnoxiously loud but contain little or no smell.

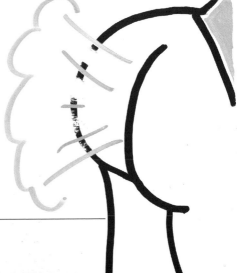

The putrid smell of some farts originates in the large intestine. After traveling through twenty-five feet of small intestine, the indigestible portions of food make their way to the colon.

Here, millions of bacteria feast on the remaining food and ferment it. And this is where things can get . . . stinky! Though the bacteria do release some useful vitamins, they also produce chemicals containing sulfur, which is responsible for the smell in flatulence.

Add to that a sulfur-rich diet with things like eggs, meat, or broccoli, and you're likely to produce some interesting odors. The longer these foods stay in the gut, the more they ferment and smell. Still, these stink bombs account for around only 1 percent of most farts.

In the absence of the odorless gases, however, farts are concentrated with smell and are generally quiet because there is less volume. Silent but deadly! Having said that, loud gas *can* be just as smelly as quiet gas—if the sulfur components are there.

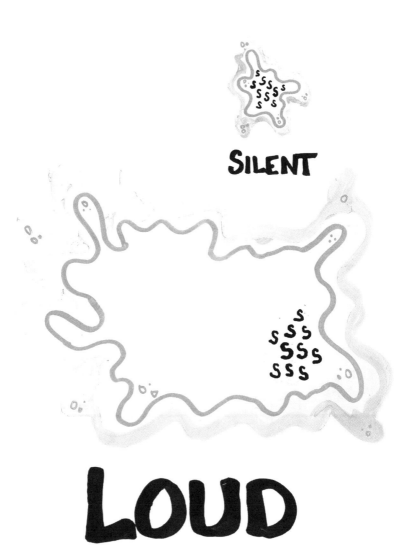

SILENT

LOUD

Simply put, your loud farts likely contain a higher proportion of air-based, odorless gas, whereas the quieter flatulence tends to have a higher proportion of the smell.

SILENT BUT DEADLY

IS CRACKING YOUR JOINTS BAD FOR YOU?

Crack! Pop! For some, the sounds bring satisfaction and relief, while for others the noise is irritating and cringe-inducing. The sharp sound of cracking joints is certainly polarizing, but how is it produced, and could cracking your knuckles really be detrimental to your health?

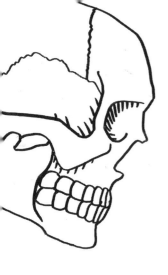

Your bones are connected by ligaments, and the areas where bones meet are your joints. In your body, there are three structurally different types of joints. Fibrous joints are fixed, like the bones in your skull; cartilaginous joints, such as in your ribs and spine, allow limited movement; and synovial joints, such as your elbows and knees, are easily movable. The synovial joints are surrounded by a special fluid that acts like grease to minimize friction and allow proper movement.

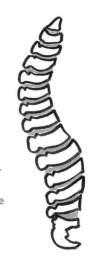

CO_2 CO_2 CO_2 CO_2 CO_2

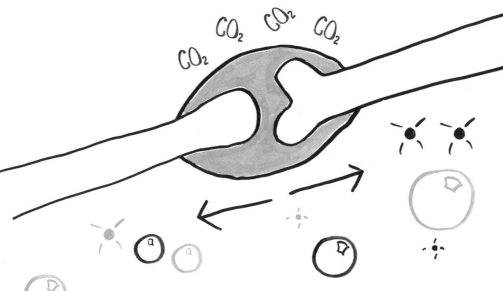

When you crack a joint, you are really stretching the joint and pulling the bones away from each other. As the bones are stretched apart, there is an increase in the volume of the joint and, consequentially, a decrease in pressure, because the physical space of the joint increases but the amount of fluid between the bones doesn't.

This lower pressure makes the gases in the fluid, such as carbon dioxide, rush to fill the newly expanded space, which starts to form bubbles. When you stretch the joints too far, you create a pressure low enough that the bubbles collapse, creating the popping/cracking noise.

The gases then take around fifteen to thirty minutes to dissolve back into the synovial fluid of the joint, which explains why you can't crack a joint again immediately after it has been cracked!

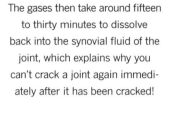

Although an urban legend suggests cracking your joints, especially your knuckles, leads to arthritis, there has been no evidence to support the claim. However, it is believed that frequent and repetitive stretching of the knuckles can cause a decrease in grip strength and some tissue damage in the joint capsules. So even though cracking your joints might feel good, overall it's probably best to let your joints be.

IS THE 5-SECOND RULE LEGITIMATE?

Whether it's the last chocolate chip cookie falling to the ground, a french fry landing on the table, or spilling open a bag of gummy bears, we've all been there asking ourselves, "Can I still eat it?" The five-second rule holds that it's okay to eat an item you dropped on the ground, as long as it was there for fewer than five seconds, right? But wait . . . is it *actually* okay?

Though there are endless variations on timing (ten-second rule, twenty-second rule, etc.), the general premise of this folklore is that your food won't be contaminated by bacteria within the given time frame. To analyze whether or not this is true, we need to understand the risks of contamination in the first place.

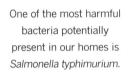

One of the most harmful bacteria potentially present in our homes is *Salmonella typhimurium*.

This particularly nasty strain of salmonella is found in the digestive tracts and feces of animals all over the world and can potentially end up in our food.

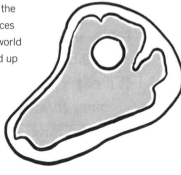

The bacteria get ingested through raw or undercooked food, and when present in large enough numbers can cause sickness. Even though the acid in your stomach will kill many bacteria, those that survive move on to the small intestine and begin to multiply, causing inflammation, which leads to cramps, diarrhea, and vomiting. So technically, you aren't "sick to your stomach"— you're sick to your *small intestine.*

While you may not be ingesting undercooked food directly, *S. typhimurium* can live up to four weeks on dry surfaces in your house (another reason to clean the kitchen more often!). Similar survival rates can be found in other bacteria, providing studies with some interesting results. A study testing the five-second rule was done by dropping bologna onto three different surfaces contaminated with *S. typhimurium*: tile, carpet, and wood.

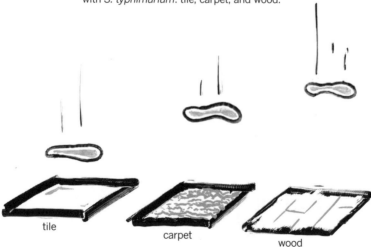

tile

carpet

wood

When the bologna was dropped onto tile, nearly 99 percent of the bacteria was transferred in five seconds! On the other hand, very little bacteria was transferred from the carpet to the bologna (0–5 percent), and a varied amount was transferred from the wood surface (5–68 percent). Carpet in the kitchen doesn't sound like such a bad idea after all!

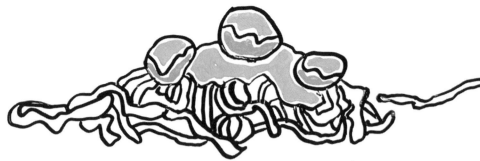

Another study found that wet food, such as pastrami, picked up much more bacteria from the surface when compared to dry food such as saltine crackers. These results remained consistent in tests using two seconds and six seconds, suggesting that it was not the amount of time that was most important but rather how wet the food was.

Finally, using a college campus to represent an "everyday environment," researchers dropped apple slices and Skittles in various dining locations to see how long it took the food to be contaminated with salmonella.

Surprisingly, the results showed that no salmonella was transferred to the dropped food, regardless of whether it was left on the ground for five, ten, or even thirty seconds. This suggests that salmonella was rarely present on the surfaces in these public spaces. Having said that, other studies have looked less specifically at one strain of bacteria and found contamination after only two seconds of contact.

So the five-second rule depends on many variables! It really comes down to *which* bacteria are present to begin with, what food you are dropping (its wetness), and which type of surface it's falling onto. Simply put, the five-second rule can be thrown out the window with regard to contamination—bacteria *will* cover your food in fractions of a second. But whether or not you will get sick depends on a variety of factors.

So knowing the risks, will you still eat it?

WHICH IS WORSE: CHILDBIRTH OR GETTING KICKED IN THE BALLS?

In the battle of the sexes, we're fiercely divided: Who has it worse when it comes to pain? On the one hand, women are faced with the prospect of fitting a watermelon-sized object through a coin-sized hole. On the other hand, males protest that even the slightest nick of their "family jewels" can leave them incapacitated. So which hurts more: childbirth or getting kicked in the balls?

There's a rumor circulating on the Internet that the human body can take up to 45 del units of pain, and yet a woman feels up to 57 del during childbirth, which is apparently equivalent to twenty bones being fractured at the same time.

The claim goes on to suggest that being kicked in the balls brings more than 9,000 del. Now, in addition to the absurd assertion that both of these events can surpass the alleged *human limit*, it uses a unit of pain, the del, that *doesn't even exist.*

DEL = FAKE

At one time, people used the term "DOL," from the Latin word for pain *dolor* to measure pain, but this has since been tossed aside in favor of other more accurate modes of assessment.

WHAT IS PAIN ?

To intelligently evaluate this question, we need to understand what pain is—which isn't an easy task. There is a group of specialized nerve cells in your body called nociceptors, which react to pain.

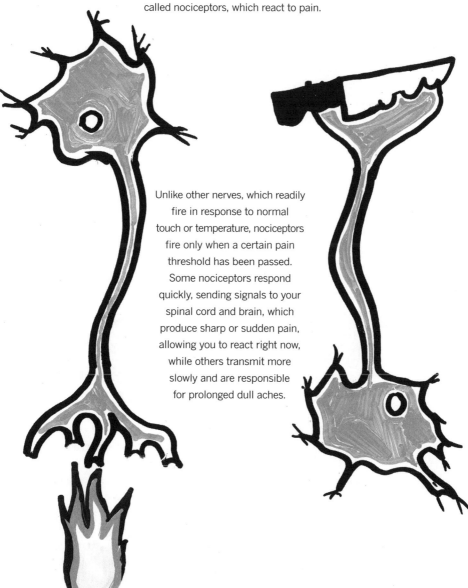

Unlike other nerves, which readily fire in response to normal touch or temperature, nociceptors fire only when a certain pain threshold has been passed. Some nociceptors respond quickly, sending signals to your spinal cord and brain, which produce sharp or sudden pain, allowing you to react right now, while others transmit more slowly and are responsible for prolonged dull aches.

MALE

VOMIT CENTER

Testicles are internal organs that have migrated out of the body cavity. And although some internal organs, such as the liver, feel no pain, others like the testicles are covered with *many* nociceptors, making them extremely sensitive. After all, their well-being is of the utmost importance.

Furthermore, the testicles are attached to many nerves in the stomach, as well as the vagus nerve, which is directly connected to the brain's "vomit center." This is why when they are hit, the pain spreads throughout your abdomen with accompanying nausea, increased blood pressure and heart rate, and sweating. Pretty bad.

FEMALE

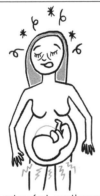

But not so fast, gentlemen. Even though childbirth may not be facing a direct hit to any internal organs, the mechanical distension (enlargement) of the uterine area also triggers nociceptors and causes the same kind of visceral pain.

Also consider that throughout evolution, female human hips have become smaller, while babies' heads have become larger—not to mention labor lasts eight hours on average, during which time women can experience nausea, fatigue, and pain. On top of it all, tension and stretching of muscle and tissue increase as labor intensifies, creating sharp and localized pain.

Okay, so both obviously hurt and have a lot of mechanical stimulation sending signals to the pain centers of the brain.

But this is where it gets tricky, because pain isn't simply a physical response but rather a partially perceptive or subjective experience. This means that every single individual perceives pain differently!

And not only between individuals, but depending on your mood, alertness, or even previous experience, pain may affect *you* differently. It's for this reason that so many attempts to objectively measure pain have failed.

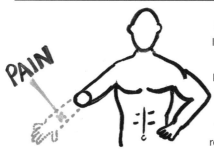

Interestingly, nearly 80 percent of upper-limb amputees experience a phenomenon known as phantom limb pain. They feel pain in a limb that is *no longer there*. And although little is understood about the mechanism for this pain, it is clear that there is no particular input to trigger the response. Yet they still feel a very real pain.

As such, pain is *not a stimulus*; it is an experience that is different for everybody. Suffice it to say, both childbirth and getting hit in the balls hurt . . . a lot. So we call this one a tie. Apart from the fact that the experiences are completely different, and there are so many variables to consider, in some instances a man could feel more pain than his female counterpart, and vice versa. The main difference between these two painful experiences is that one results in a newborn baby, while the other potentially results in a decreased chance of having one.

PAIN
IS
SUBJECTIVE

DOES SHAVING MAKE YOUR HAIR GROW THICKER?

We've all been cautioned at one point or another that once you shave, your hair will grow back thicker or darker than it was before. In fact, some say it even grows faster after shaving. Is there any truth to these warnings?

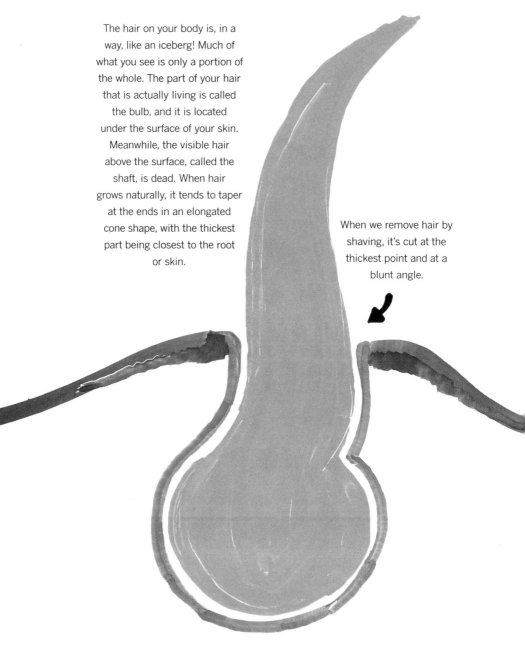

The hair on your body is, in a way, like an iceberg! Much of what you see is only a portion of the whole. The part of your hair that is actually living is called the bulb, and it is located under the surface of your skin. Meanwhile, the visible hair above the surface, called the shaft, is dead. When hair grows naturally, it tends to taper at the ends in an elongated cone shape, with the thickest part being closest to the root or skin.

When we remove hair by shaving, it's cut at the thickest point and at a blunt angle.

So as the hair grows back, and the new shaft pushes through the skin, rough stubble emerges. But this stubble is the exact same thickness as the hair you previously chopped off.

While the stubble may *appear* slightly thicker, in truth, it really isn't. The size of your follicle determines the thickness of the hair, and shaving the dead hair shafts can't change that.

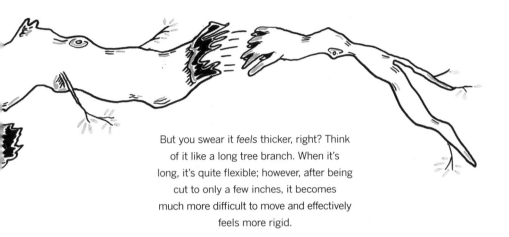

But you swear it *feels* thicker, right? Think
of it like a long tree branch. When it's
long, it's quite flexible; however, after being
cut to only a few inches, it becomes
much more difficult to move and effectively
feels more rigid.

How about the color of hair? This is determined by the cells in your skin that
produce melanin (the pigment that gives your skin and hair color),
called melanocytes. Hair growth after shaving sometimes appears to be darker,
but the color of the hair is truly unaltered. Stubble may be more noticeable
for a couple reasons. First, seeing the short hair dots directly against the
backdrop of your skin creates greater contrast, and second, this hair has not
yet had the chance to be lightened by sun or chemical exposure.

When it comes to the speed at which your hair grows, again, the myth is busted. All studies done on hair growth rate have concluded that there is absolutely no effect from shaving.

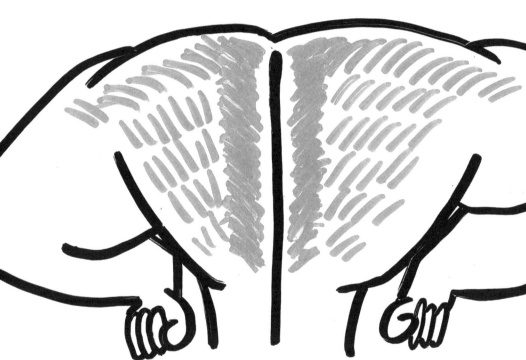

So next time you reach for your razor, don't worry about those old wives' tales! Chop that ugly, unwanted hair off your back, because science has got *your* back.

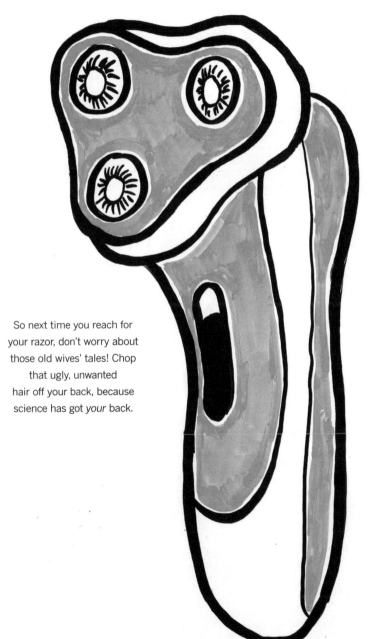

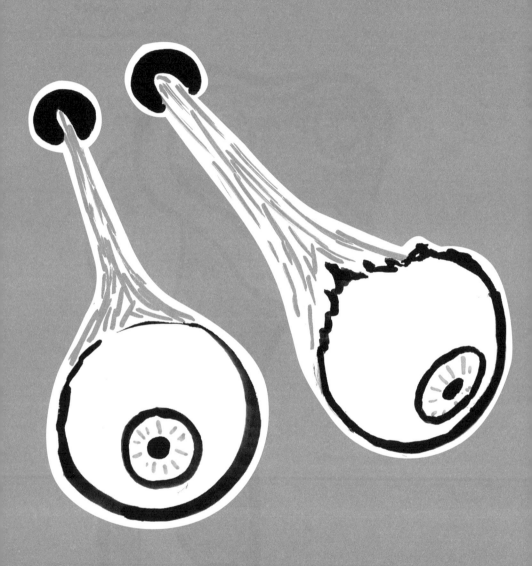

CAN SNEEZING POP YOUR EYEBALLS OUT?

More often than not, it comes from out of the blue. With only a second's notice to brace yourself—*achoo!* Did your heart stop beating? Have your eyeballs popped out? With such a forceful reaction one has to wonder: What causes this outburst, and how powerful is it, really?

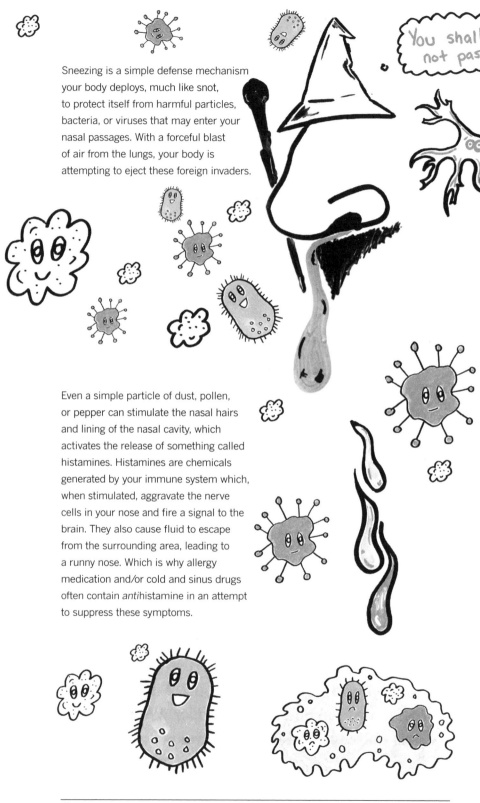

Sneezing is a simple defense mechanism
your body deploys, much like snot,
to protect itself from harmful particles,
bacteria, or viruses that may enter your
nasal passages. With a forceful blast
of air from the lungs, your body is
attempting to eject these foreign invaders.

You shall
not pass!

Even a simple particle of dust, pollen,
or pepper can stimulate the nasal hairs
and lining of the nasal cavity, which
activates the release of something called
histamines. Histamines are chemicals
generated by your immune system which,
when stimulated, aggravate the nerve
cells in your nose and fire a signal to the
brain. They also cause fluid to escape
from the surrounding area, leading to
a runny nose. Which is why allergy
medication and/or cold and sinus drugs
often contain antihistamine in an attempt
to suppress these symptoms.

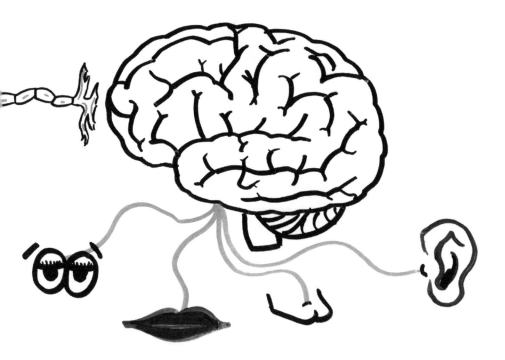

The signal sent to your brain travels through the trigeminal nerve network, which controls most parts of your face—eyes and eyelids, forehead, scalp, cheeks, teeth, chin, jaw, and even your outer ear. This amazing network of nerves prompts the "sneeze center" in the lower brain to set a sneeze in motion. Your brain quickly communicates the signal to a team of muscle groups in your face, throat, and chest, which reflexively respond and work together to produce a sneeze, eliminating whatever is irritating your nasal cavity.

So how powerful is it? These gargantuan gusts of wind travel around 30 to 40 mph on average!

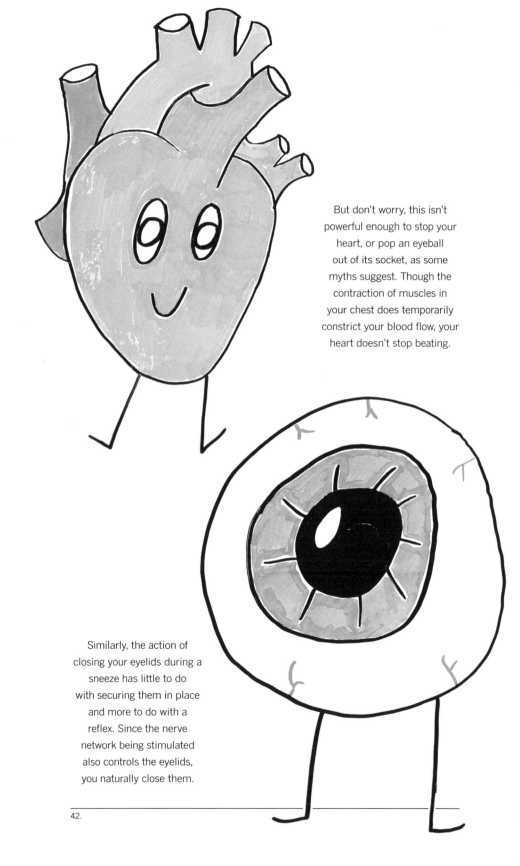

But don't worry, this isn't powerful enough to stop your heart, or pop an eyeball out of its socket, as some myths suggest. Though the contraction of muscles in your chest does temporarily constrict your blood flow, your heart doesn't stop beating.

Similarly, the action of closing your eyelids during a sneeze has little to do with securing them in place and more to do with a reflex. Since the nerve network being stimulated also controls the eyelids, you naturally close them.

But what about when you sneeze at extremely peculiar times? Whether after a quick drop in temperature, following exercise, sex, or on a full stomach, many people suffer sneeze attacks triggered by unusual sources. Sudden exposure to bright light, also known as the photic sneeze reflex or ACHOO (Autosomal-dominant Compelling Helio-Ophthalmic Outburst) syndrome, is thought to affect around a third of the population.

It's believed that a signal from the optic nerve somehow crosses the trigeminal nerve. So instead of informing the brain that light is irritating the eye, a biological error is made and the trigeminal nerve informs the brain that the nose is irritated, resulting in a sneeze!

So although a sneeze can sometimes feel like a spontaneous and unintentional full-body workout, it's just your body's way of cleansing your nose the best way it . . . knows.

COULD YOU
SPONTANEOUSLY
COMBUST?

Is it possible for a human body to suddenly burst into flames? The idea may seem absurd, and yet spontaneous human combustion has been reported hundreds of times since its first documented case in 1663. But how real are these cases, and is there any science behind them? More important, could it happen to *you*?

Most spontaneous human combustion cases follow a similar story:
An elderly individual, often overweight, is found burnt to a
crisp after being alone at home. And although the surroundings
show some evidence of fire, the home is generally untouched.

The most peculiar aspect is that the victim's extremities—the hands and
feet—often remain intact. But perhaps the most important aspect
is that, in almost every case, these fires are never actually
seen to be spontaneous. Instead, only the aftermath is witnessed.

THE WICK EFFECT

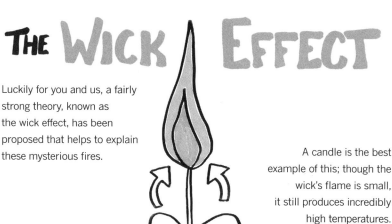

Luckily for you and us, a fairly
strong theory, known as
the wick effect, has been
proposed that helps to explain
these mysterious fires.

A candle is the best
example of this; though the
wick's flame is small,
it still produces incredibly
high temperatures.
As a result, it melts the
surrounding wax, which
is drawn up the wick
and vaporized. This vapor-
ization cools the wick and
prevents it from burning,
until the wax is gone, at
which point the wick burns
to a crisp.

How is this related to human combustion? The theory suggests that a cigarette or other small flame source could burn a small area of clothing, which in turn burns the skin underneath.

Eventually this skin splits open and releases fat, which surrounds the burning clothes much like wax on a candle. (In fact, candles were originally made with animal fat.) With the clothes acting as a wick and the fat as wax, the small fire will burn slowly, without spreading to the surroundings, as long as the fuel source is available.

This has been tested on pigs and helps explain why the hands and feet are sometimes left unaffected, as these body parts have low levels of fat.

Even though this hypothesis starts with an external ignition source (as opposed to being truly spontaneous), researchers have good reason to believe it's correct. After looking at cases in the eighteenth, nineteenth, and twentieth centuries, a published investigation concluded that most victims were near a plausible source of ignition: a candle, a fireplace, cigarettes. The majority of victims were also elderly or those with low mobility, making an escape much more difficult. Finally, evidence suggests a good chance that many died in their sleep.

Still think it may be real? Consider that no other animal on the planet has ever been documented to spontaneously combust. The phenomenon is uniquely human.

So sleep tight knowing you won't likely spontaneously combust!

IS BINGE TV WATCHING BAD FOR YOU?

Whether it's watching the Olympics for two weeks straight, playing a new video game start to finish, or bingeing on your favorite TV show, most of us have spent a significant amount of time in front of a screen. So what effect does this extended watching, or bingeing, have on the body?

In the past, it may have actually been deadly! In 1967, an error in manufacturing caused some TVs to emit harmful X-rays, with radiation levels one hundred thousand times higher than what is considered safe today.

But even modern televisions cause a strain on your eyes. Under normal circumstances, humans blink around eighteen times per minute, but when staring at a screen this rate decreases drastically, causing sore and tired eyes. Fortunately, these symptoms are generally short term.

For children, however, simply spending extended hours *indoors* can have developmental effects. A condition called myopia—where the eyes cannot focus properly—is seen much more frequently in children who spend a lot of time inside. Not only are they constantly focusing on nearby objects, as opposed to far-off landscapes and distances, but scientists believe that the sun itself may actually play a role in healthy eye regulation.

And while TV may feel like a nice way to relax your body and brain, that may not always be a good thing. Not only is a sedentary lifestyle a major contributor to obesity, but studies have also shown that people who watch less TV tend to burn more calories—even if they aren't doing more physical activity! Simply doing more mentally rigorous tasks like reading, playing a board game, or simple household activities requires more energy and burns more calories. Look at that! You're already partway there just by reading this book.

And if you truly want to relax, TV before bed may also be hurting you. Studies have shown that it may actually reduce your hours of quality sleep, contributing to chronic sleep debt. It may also affect other bedtime activities; researchers have found that men who watch more than twenty hours of TV a week have, on average, a 44 percent reduction in sperm.

But perhaps the most significant findings relate directly to your life span. Not only is there a documented correlation between TV viewing time and the risk of diabetes and heart disease, but shockingly, multiple studies have also found a correlation between TV viewing time and *all causes of death.* One study concluded that every hour spent in front of the TV may cut as much as twenty-two minutes off your life.

Of course, correlation does not equal causation—you can have a healthy relationship with TV if consumed in moderation. At the root of many of these findings is that the physical inactivity associated with prolonged TV watching is harmful. Ultimately, the more you move, the more you live!

— WHICH CAME FIRST —
THE CHICKEN OR THE EGG?

It's a question that has perplexed humanity from as early as the ancient Greeks all the way to the twenty-first century, and we're still dying to know: Which came first—the chicken or the egg?

The question would be simple if we took it literally. Egg-laying animals existed around 340 million years ago, long before chickens came about, so technically the egg came before the chicken.

But this question, better worded as "the chicken or the *chicken* egg," focuses more on the cyclical cause and consequence; that is, if a chicken is born from an egg, where did the egg come from? Another chicken, presumably, which too must have come from an egg. So which came *first*?

TEAM CHICKEN

On one side, we have Team Chicken. Research suggests that the protein essential for the formation of chicken eggs, called OV-17, is found *only* in chicken ovaries. Without it, the chicken eggshell cannot be formed. So without a chicken, you technically can't get a chicken egg.

But this all depends on the nature and definition of a chicken "egg" in the first place. After all, is a chicken egg an egg laid *by* a chicken, or one that simply *contains* a chicken? Obviously the OV-17-bearing chicken had to come from somewhere. But if an elephant laid an egg from which a lion hatched, would it be an elephant egg or a lion egg?

TEAM EGG

This leads to the other side of the story: Team Egg. During reproduction, two organisms pass along their genetic information in the form of DNA.

But the replication of this DNA is never 100 percent accurate and often produces minor changes for the new organism. These small mutations in DNA, over thousands of generations, create new species. But these genetic mutations must occur in the zygote, or the initial cell—the fertilized egg.

So a creature very similar to a chicken, let's call it a protochicken, would have mated with another protochicken, and because of a small genetic mutation created the first chicken— which grew in an egg.

MUTATION

So the egg came first? Well, Team Chicken might argue that this was simply a chicken growing inside of a protochicken egg.

However, no one mutation can ever really constitute a new species. Even though we humans like to classify all creatures into different groups with different names, this is based on how things currently are and not how they were millions of years ago. The process of evolution is so gradual that no one protochicken-to-chicken birth could be considered a new species *at the time*. Much like how dogs have evolved from wolves. As humans began to interact with and domesticate wolves, there was no one single point where a wolf gave birth to a dog.

Rather, as particular traits came about from selective pressures—
such as choosing wolves that were not afraid of humans,
or ones that were less aggressive—over many generations we can see
big genetic and behavioral trait differences.

So we're left with two scenarios:

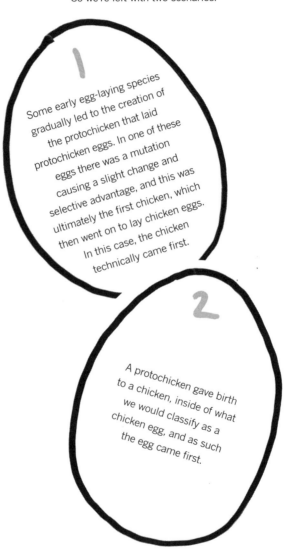

1

Some early egg-laying species gradually led to the creation of the protochicken that laid protochicken eggs. In one of these eggs there was a mutation causing a slight change and this was selective advantage, and this was ultimately the first chicken, which then went on to lay chicken eggs. In this case, the chicken technically came first.

2

A protochicken gave birth to a chicken, inside of what we would classify as a chicken egg, and as such the egg came first.

Which brings us back to the nomenclature and question of
"What is a chicken egg?" which is a fairly meaningless question.
But at the end of the day, what we *can* all agree on is that,
regardless of whether it was a chicken egg or a protochicken
egg, the first true chicken came from an egg.

THE EGG CAME FIRST.

BODY

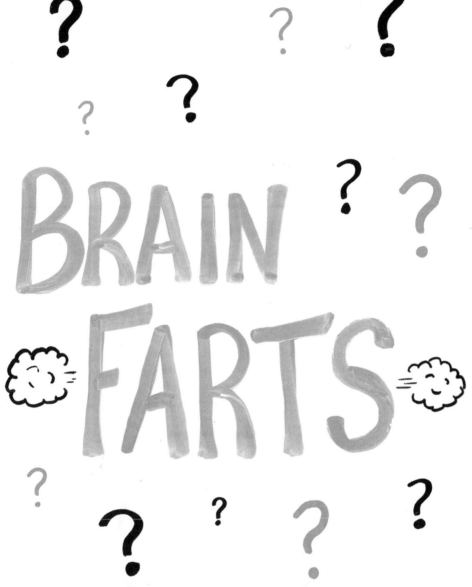

BRAIN FARTS

They may not smell, but they definitely stink.
You know, those times when your brain seemingly
forgets how to function or you're unable to
speak like a normal human being for about five
seconds? Oops—your brain just farted! But
what *really* happened?

The scientific term for this phenomenon is "maladaptive brain activity change." Yes, scientists have actually devoted time to understanding this conundrum.

MALADAPTIVE BRAIN ACTIVITY

After monitoring brain activity in individuals during repetitive tasks, they began to notice something incredible: They could actually see abnormal brain activity up to thirty seconds *before* the mistake was made.

This was a big surprise, because many assumed that these blunders were simply caused by a momentary lapse in concentration.

Instead, almost half a minute before an error,
the brain regions associated with relaxation
become active, while those linked to maintaining
task effort begin to shut down.

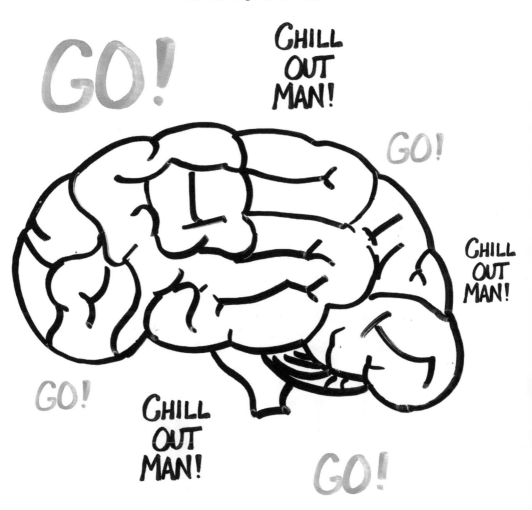

But as soon as you notice your mistake or lapse, brain activity
kicks into overdrive and goes back to normal. These types
of mistakes are much more common during repetitive or overly
familiar activities. Scientists believe this is the brain's attempt
to save energy during a task, by entering a more restful state.
However, sometimes the brain takes the relaxation a bit too far,
leading to your slipup.

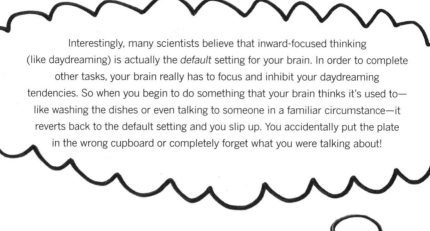

Interestingly, many scientists believe that inward-focused thinking (like daydreaming) is actually the *default* setting for your brain. In order to complete other tasks, your brain really has to focus and inhibit your daydreaming tendencies. So when you begin to do something that your brain thinks it's used to—like washing the dishes or even talking to someone in a familiar circumstance—it reverts back to the default setting and you slip up. You accidentally put the plate in the wrong cupboard or completely forget what you were talking about!

But not to worry—it's a completely
normal part of human life
that most of us have experienced,
and will continue to. As with
regular farts, you'll just have to
learn to live with them!

THE SCIENCE OF BAD BREATH

Whether it's the stinky aroma from your garlic-infused dinner, or a strong blast of it first thing in the morning, we've all had our fair share of bad breath. But how does that smelly scent develop, and why do some people have it far worse than others? More important, how can we get rid of the dreaded stench?

Bad breath is a common condition experienced by people of varying ages and genders, and is present in nearly 50 percent of the population at any one time. Meanwhile, nearly 20 percent of the population suffers from chronic bad breath, otherwise known as halitosis. Think you may have it? Not so fast. The vast majority of people who believe they have halitosis actually suffer from halitophobia—the fear of bad breath.

For the bulk of the population, bad breath can be attributed to our dear friends bacteria.

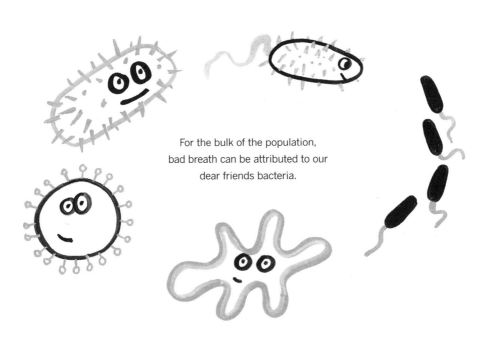

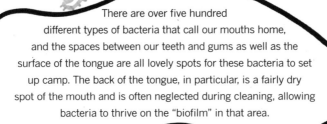

There are over five hundred different types of bacteria that call our mouths home, and the spaces between our teeth and gums as well as the surface of the tongue are all lovely spots for these bacteria to set up camp. The back of the tongue, in particular, is a fairly dry spot of the mouth and is often neglected during cleaning, allowing bacteria to thrive on the "biofilm" in that area.

These bacteria break down food remnants and release volatile sulfur compounds, known for their rotten-egg smell.

Just like when you leave food or milk out on the counter too long, bacteria start to break down the proteins and produce an offensive or sour smell.

It may seem gross, but it's actually the same product bacteria release when breaking down food in your intestines, ultimately producing the smell of farts. Meaning when you have bad breath, you're really just farting out of your mouth . . .

Of course, certain foods stimulate bacterial growth more than others. Foods dense with protein like dairy, meat, and fish can break down into volatile sulfur compounds.

Mouth-drying agents, like alcohol and cigarettes, create an ideal environment for the bacteria to flourish in.

Coffee provides an acidic and potentially sugar-rich environment, both of which increase bacteria reproduction and lead to the infamous "coffee breath."

And while many of these can be managed with proper oral hygiene and hydration, certain odorous foods like onions and garlic already contain sulfur compounds, which are released into your bloodstream. These then travel to your lungs and the pores of your body, only to be radiated or breathed out.

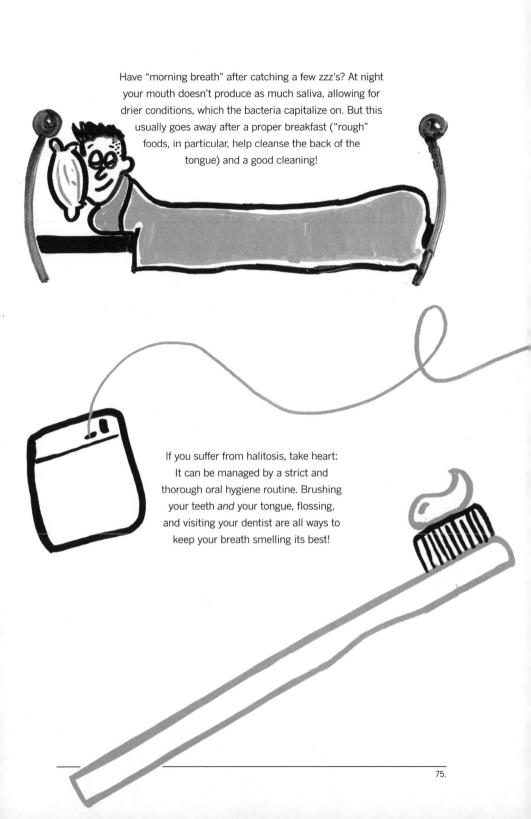

Have "morning breath" after catching a few zzz's? At night your mouth doesn't produce as much saliva, allowing for drier conditions, which the bacteria capitalize on. But this usually goes away after a proper breakfast ("rough" foods, in particular, help cleanse the back of the tongue) and a good cleaning!

If you suffer from halitosis, take heart: It can be managed by a strict and thorough oral hygiene routine. Brushing your teeth *and* your tongue, flossing, and visiting your dentist are all ways to keep your breath smelling its best!

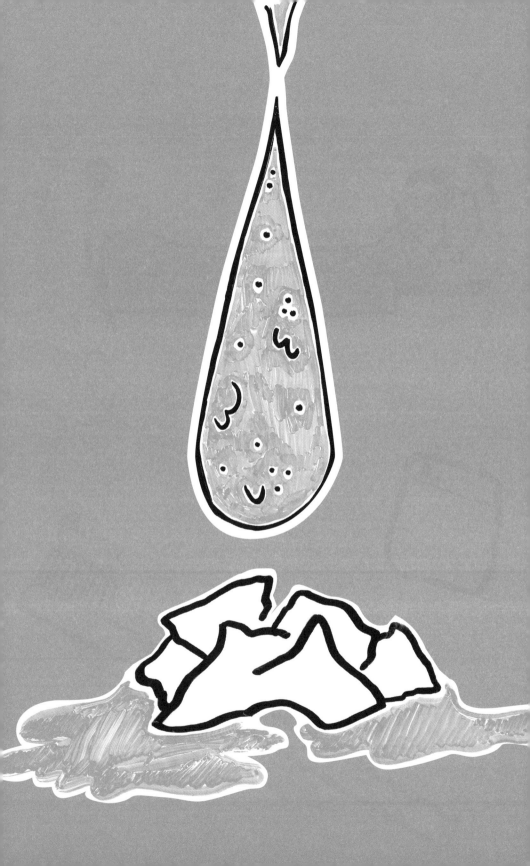

CAN SNOT SAVE US?

Seeing someone blow his nose or eat a booger is gross—disgusting, even! But is snot really all that bad? What if we told you it could actually save your life, and that this sticky substance should be revered instead of feared?

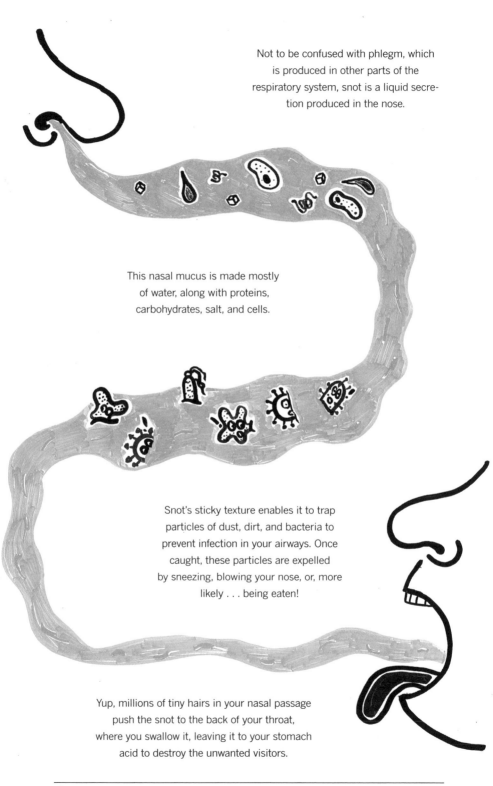

Not to be confused with phlegm, which is produced in other parts of the respiratory system, snot is a liquid secretion produced in the nose.

This nasal mucus is made mostly of water, along with proteins, carbohydrates, salt, and cells.

Snot's sticky texture enables it to trap particles of dust, dirt, and bacteria to prevent infection in your airways. Once caught, these particles are expelled by sneezing, blowing your nose, or, more likely . . . being eaten!

Yup, millions of tiny hairs in your nasal passage push the snot to the back of your throat, where you swallow it, leaving it to your stomach acid to destroy the unwanted visitors.

Fancy a taste? You may have noticed a lot of kids picking their noses and chowing down, but surely it's bad for them, right? Not quite! Researchers have theorized that nasal mucus may contain a sugary taste to entice young people to eat it. In a society devoid of dirt and germs, and increasing allergies and disease, eating boogers may actually be a way for children to expose themselves to pathogens, which may ultimately help build up their immune system.

A healthy nose pumps out around half a liter of snot a day. If you have a cold, a virus infects your body and moves into the mucous membranes of your cells and multiplies. Your body responds by inflaming the mucous cells, pumping more blood to your nose and leaking more water through the cells—hence your runny nose. Simultaneously, your body sends white blood cells to attack the virus with potent chemicals or engulf it entirely.

Along with the antiseptic enzymes in snot, which directly kill bacteria, there is an abundance of proteins called mucins, which are designed to prevent bacterial growth.

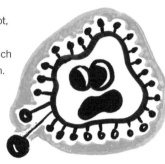

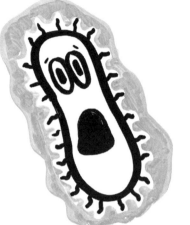

Mucins have a dense sugar coating that allows them to hold water and create a gel-like consistency, which keeps bacteria from moving around and clumping together. By separating individual bacterial cells, they can't work together. As scientists begin to look more closely at these mucins, there is a potential for use in products like toothpaste and even hospital surfaces, where large amounts of bacteria grow together.

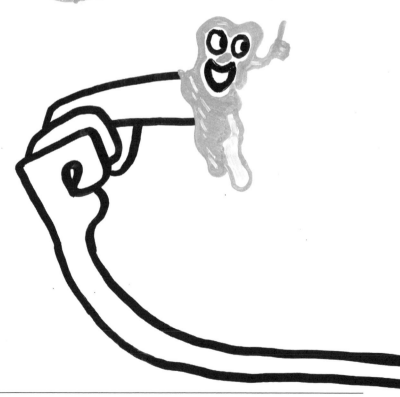

Notice color in your snot? It can tell you a lot! Small amounts of red blood generally mean too much rubbing, blowing, or picking, while green snot generally indicates a viral or bacterial infection. The green color is part of your immune response and comes from iron present in snot enzymes. In fact, these are the same enzymes that create the green color of wasabi, which was originally used in Japanese cuisine to combat bacterial contamination of raw foods.

Clear snot, on the other hand, generally means your nose is healthy.

So raise a finger in honor of our gooey friend snot, which in spite of all the flack, has always got your back!

IS THERE A HICCUP CURE?

"Drink a glass of water upside down. It works every time." "Just hold your breath and count to ten." "Bite a lemon." Some people swear by these remedies, but more often than not we end up waiting the hiccups out. So what are hiccups, and is there any surefire way to get rid of them?

Hiccups are directly related to your diaphragm, a
muscle that stretches across the bottom of
your rib cage and plays a central role in breathing.

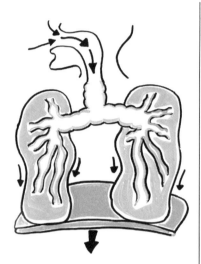

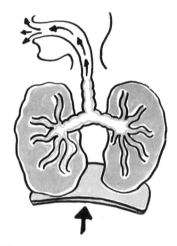

When you inhale, your diaphragm
contracts and is pushed
downward, creating more space
for your lungs to expand.

When your diaphragm relaxes, it moves
back upward, forcing your lungs to
shrink and exhale. These movements
happen with no conscious effort,
day in and day out. However, there are
times when this repetitive system
can be thrown off track—with a literal
hiccup in the system!

A hiccup is simply a short, involuntary twitch of the diaphragm that triggers
a contraction of the muscle and a closing of the vocal cords, which
produces the classic *hic* sound. While it's possible to experience a lone hiccup,
we often endure them in succession, at a rhythmic rate ranging from
four to sixty per minute.

4-60 PER MIN

But why? Your nerves send signals between your body and brain, and occasionally there is a disruption in this pathway. As a result, the signal is interrupted and the automated breathing action becomes irregular. Where exactly the disruption is happening remains unclear.

Hiccups can be brought on by a variety of factors but are often associated with consuming carbonated or alcoholic drinks and especially eating fatty or spicy foods.

The spicy or fatty foods can disturb the nerves of the stomach and diaphragm, creating a hiccup.

WHAT'S THE CURE?

So is there a magic cure? Well, of all of the many proposed methods, none have been scientifically studied and experimentally proven, except one. And you're probably not going to like it.

It turns out that in two independent studies, researchers found that a rectal massage stopped the hiccups within thirty seconds. That's right—sticking a finger up the anus and wiggling it around. It just so happens to stimulate some of the same nerves involved with hiccups, which potentially jolts the system back to normal. Granted, in both documented cases, the patients were suffering from persistent hiccups, not just normal, short-term episodes. But both studies came to this same conclusion separately, giving it some weight. On the other hand, while not tested, an orgasm can stimulate these same nerves too . . . which you may prefer to rectal stimulation. Your body—your decision!

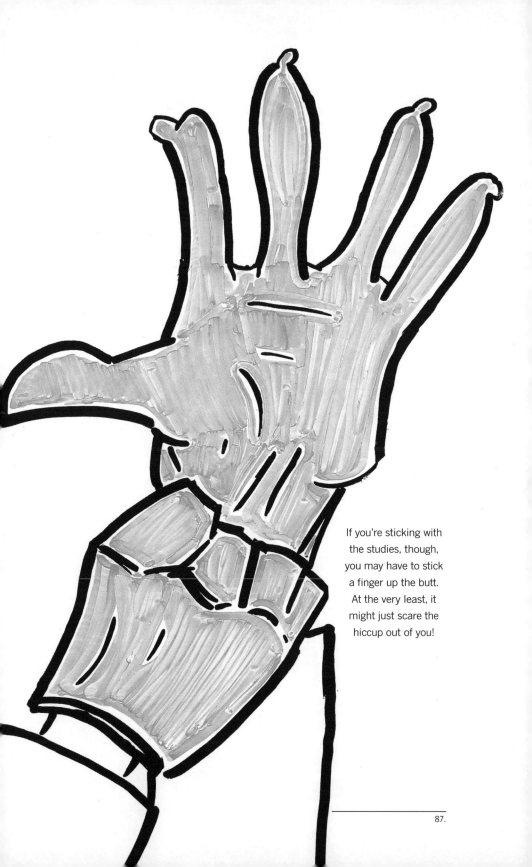

If you're sticking with the studies, though, you may have to stick a finger up the butt. At the very least, it might just scare the hiccup out of you!

THE SCIENTIFIC SECRET OF

MUSCLE & STRENGTH

So many of us are fascinated by muscles and strength—whether we want to be stronger, *feel* better, or look like Superman; and there are many proposed strategies and exercise regimens to get this done. But what if we told you that it was out of your control, and that your genetics may be holding you back entirely? Conversely, what if there was a secret out there that could lead you to superhuman strength?

The truth is: Your muscle size has a limit. Sure,
this may seem obvious when you hear it,
but your muscles are under the strict control of a
protein called myostatin, which determines
exactly how large a muscle can *become*.

MYOSTATIN

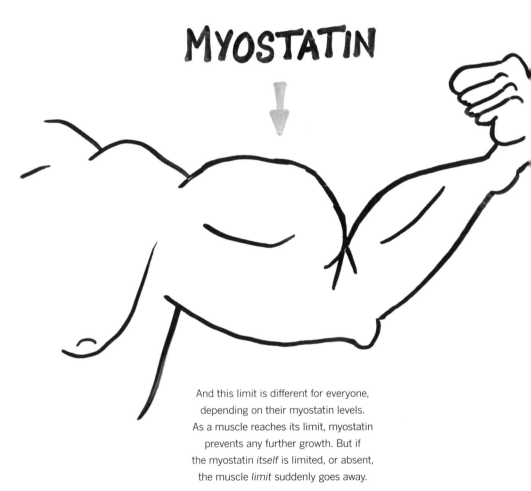

And this limit is different for everyone,
depending on their myostatin levels.
As a muscle reaches its limit, myostatin
prevents any further growth. But if
the myostatin *itself* is limited, or absent,
the muscle *limit* suddenly goes away.

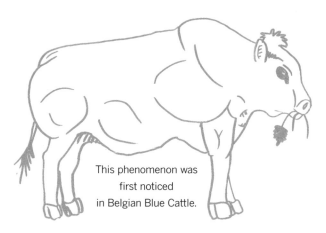

This phenomenon was
first noticed
in Belgian Blue Cattle.

These cows developed two- to three-times more muscle mass than a
normal cow, and it was later discovered that they had a deletion
of the gene GDF-8, which just so happens to create myostatin. As a
result, without any exercise or special diet, these cows have incredible
muscle mass. Similar cases have been documented in dogs, mice,
and even a few cases of human babies lacking the GDF-8 gene.

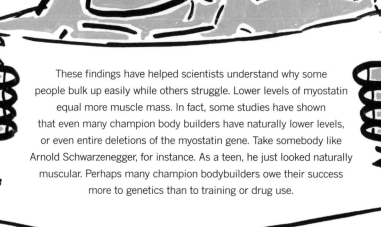

These findings have helped scientists understand why some
people bulk up easily while others struggle. Lower levels of myostatin
equal more muscle mass. In fact, some studies have shown
that even many champion body builders have naturally lower levels,
or even entire deletions of the myostatin gene. Take somebody like
Arnold Schwarzenegger, for instance. As a teen, he just looked naturally
muscular. Perhaps many champion bodybuilders owe their success
more to genetics than to training or drug use.

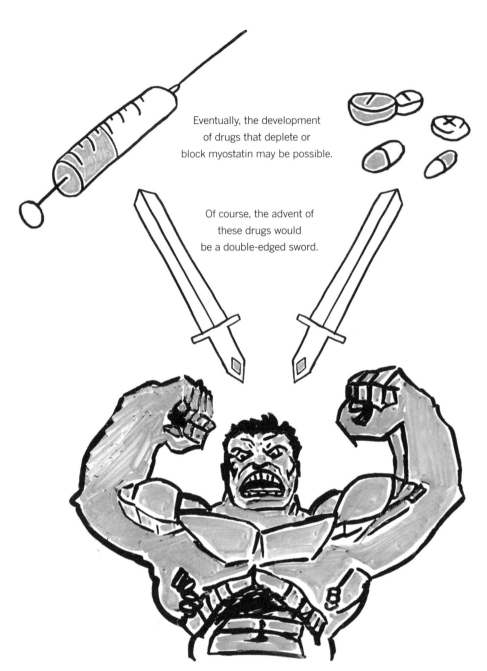

Eventually, the development of drugs that deplete or block myostatin may be possible.

Of course, the advent of these drugs would be a double-edged sword.

On the one hand, patients with muscular dystrophy or muscle loss from aging could regrow and restore muscle. On the other hand, the potential for the drugs' abuse as performance enhancers seems inevitable. But the prospect of seeing a real-life human Hulk seems kind of cool. Besides, should those who already have low levels of myostatin be considered to have an unfair advantage? It's hard to say.

Regardless, be wary of anybody
selling or promoting this unproven
and unregulated treatment.

At the end of the day, whether you
have naturally low or high myostatin
levels, remember: This is the level that is
optimal for *your* mechanical and metabolic
efficiency. With proper exercise and general health,
you'll reach your own unique physiological peak.
And while you may not be destined to pack a punch
like Superman, you *can* push your own limits
through training and hard work.

If you want to know how much myostatin *you* have,
there's only one way to find out. Start working!

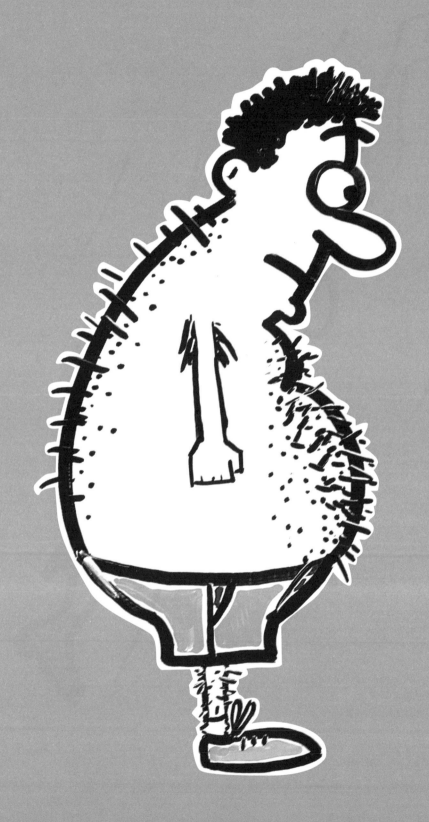

WHY ARE MEN SO HAIRY?

Hairy heads, hairy legs, and hairy armpits—the human body is covered in the stuff! And although some of us attempt to get rid of hair in "weird" places, it is still obvious that men have more body hair than women. So what is the point of body hair, and why are men so hairy?

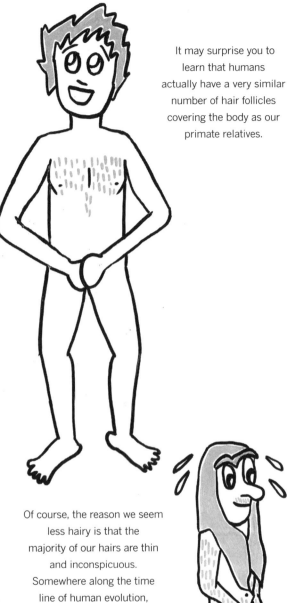

It may surprise you to learn that humans actually have a very similar number of hair follicles covering the body as our primate relatives.

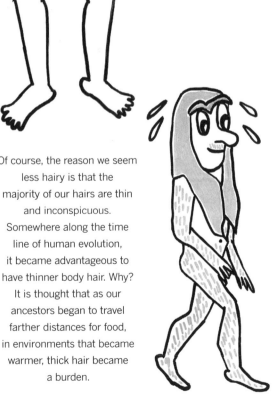

Of course, the reason we seem less hairy is that the majority of our hairs are thin and inconspicuous. Somewhere along the time line of human evolution, it became advantageous to have thinner body hair. Why? It is thought that as our ancestors began to travel farther distances for food, in environments that became warmer, thick hair became a burden.

To survive in sweltering heat with increased activity,
it became biologically advantageous to have thin body hair
and more sweat glands. This allows heat to leave easily
through the body, thus cooling it; however, hair remained
on the head to protect the brain from direct sunlight.

As we share a similar number of follicles with primates,
men and women share roughly the same amount
as well. However, men have more terminal hair—the longer,
thick, and colored hair most commonly associated
with head hair and pubic hair. On men, it can also grow on
the chest, back, toes, in the ears, and a variety of
other odd places. Women, on the other hand, have more
vellus hair, which is thinner and less visible.

Why the difference? Evolution has likely driven men to become hairier through the mechanism of sexual selection. Animals of the opposite sex determine each other's fitness, health, and fertility based on external traits. Male peacocks, for example, use their flashy tail feathers to display their fitness and attract females.

In a similar way, a hairy chest may have been selected to indicate health and entice female mates. Another theory suggests that men may have preferred women with less hair, as it makes them look younger, alluding to their healthy childbearing years.

Recently it has been discovered that hairy men are better at detecting parasites on their body (like bedbugs) than are nonhairy men. From this, theories have developed suggesting that females were attracted to hairy men because it meant they would be healthier, parasite-free sexual partners.

Even among men, hairiness varies widely depending on ethnic groups and genetics. Some men choose to embrace their hairiness while others shave it off. Contemporary cultural trends get replaced too quickly to create a persuasive evolutionary pressure, so feel free to shave, wax, pluck, and design your hair any way you like.

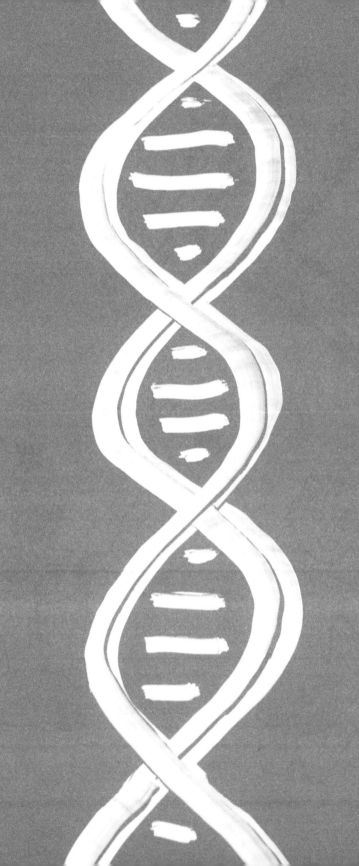

THE SCIENCE OF AGING

While many search for the elusive Fountain of Youth, you might be wondering, Why do we age in the first place? What is it about our bodies or cells, biologically, that causes us to grow old?

A variety of internal and external factors such as diet, exercise, and environmental stress contribute to cell damage and repair, and affect the rate of aging.

WE'RE PROGRAMMED TO DIE

But the surprising truth is that, apart from these, we actually have a biological clock buried within our genetic makeup. And this clock can only run for so long—in other words, we're *programmed* to die.

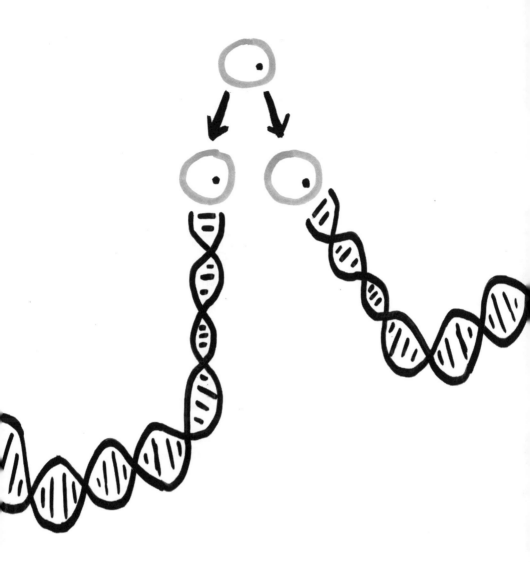

Your body is made up of trillions of cells, which are constantly going through cell division. And every time they divide, they make a copy of their DNA. This DNA is tightly packed into structures called chromosomes, of which humans have twenty-three pairs.

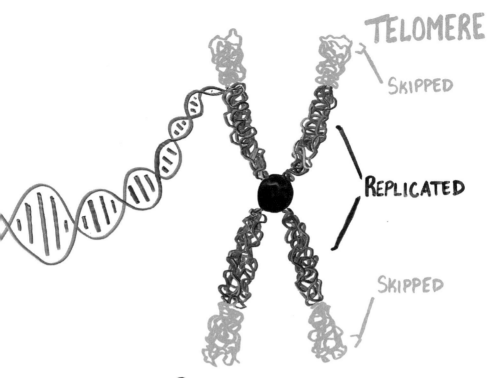

TELOMERE

SKIPPED

REPLICATED

SKIPPED

CHROMOSOME

The problem is, DNA replication isn't perfect and skips over the end of each chromosome. To protect important DNA information from being cut out, we have something called *telomeres* on the ends of chromosomes. These are essentially meaningless repeats of DNA that we can afford to lose.

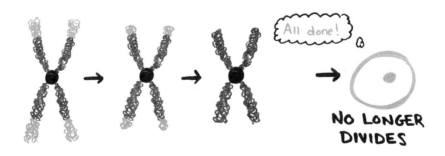

All done!

NO LONGER DIVIDES

But every time your cells divide, the telomeres become shorter and shorter, until eventually they've been entirely stripped away. At which point, the cell no longer divides.

Some flatworms are able to endlessly regenerate their telomeres, making them effectively biologically immortal. But their life spans do vary, and they're still susceptible to disease, further suggesting that aging is a mix of genetic and environmental factors.

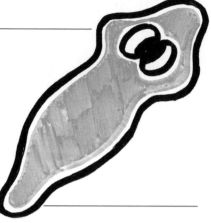

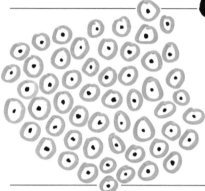

Why don't our cells do this? Ultimately, the replication limit actually helps to prevent cancer, which is the uncontrollable growth of cells and evasion of cell death.

The point at which a cell stops replicating is known as cellular senescence. In humans, this replication limit is around fifty times.

CELLULAR SENESCENCE
HUMANS ≈ 50

Once it's reached, the cell gradually begins to lose its function and dies, causing age-related characteristics. This also helps to explain why life expectancy is a strongly heritable trait from your parents—because you got your initial telomere length from them.

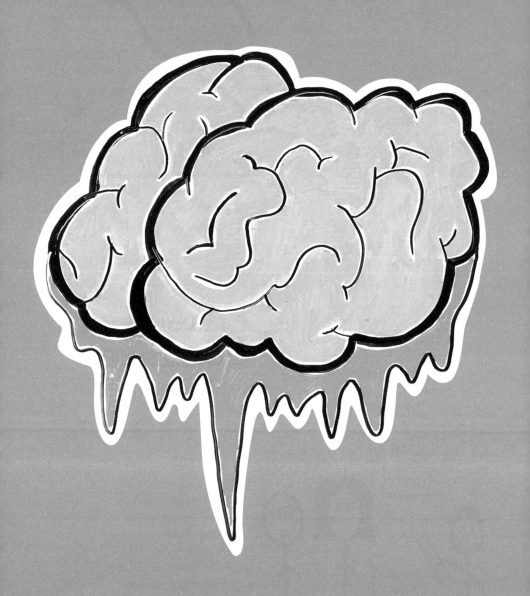

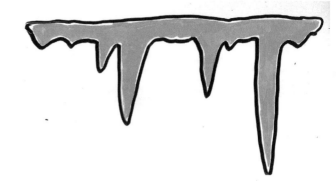

THE SCIENCE OF

BRAIN FREEZE

There's nothing quite worse than the anticipation of pure bliss as a refreshing cold beverage or ice cream hits your mouth, only to have your brain freeze up with pain. So what's going on up there during brain freeze? And what can it tell us about how the brain works?

Typically, brain freeze pain or headaches last around twenty seconds and are triggered when a cold substance touches the roof of your mouth. Similar to when you're in a cold environment, the smallest blood vessels in your body, called capillaries, constrict to conserve heat.

Many of these capillaries are located in your extremities, and so by constricting they allow heat and blood to be conserved in the core of your body, where your vital organs are.

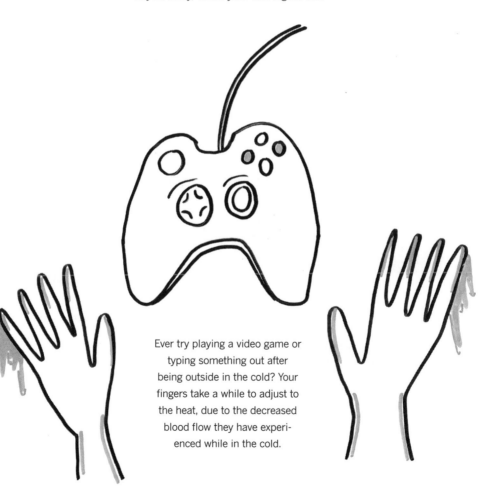

Ever try playing a video game or typing something out after being outside in the cold? Your fingers take a while to adjust to the heat, due to the decreased blood flow they have experienced while in the cold.

The same thing happens to the capillaries in the roof of your mouth.

Capillaries Constrict

But in this case it's not constriction that causes the pain. Instead, it's the fact that more blood heads to your brain in an effort to keep it warm.

And because your brain is contained in your skull, all the extra blood causes an increase in pressure, leading to a headache. The brain is perhaps the most important organ in your body and has developed extremely fast artery dilation and constriction methods as protection. And so when faced with extreme cold—whether outside or inside the body—it reacts immediately.

Brain Swells

Skull Causes Pressure

But once the substance has been removed or swallowed, the capillaries in your mouth rapidly dilate (expand), potentially causing even more pain.

HEADACHE!

The same nerve that senses pain in the forehead has receptors on the roof of your mouth that detect the dilated capillaries, and these receptors send a pain signal to your brain. Which is why your forehead often feels the bulk of the pain.

Scientists have studied brain freeze as a way to learn more about the physiology of regular headaches as well. Unfortunately, there is no other reliable (or ethical) way to induce a headache or migraine in a lab setting. So it was by giving volunteers cold beverages to induce brain freeze that they were able to see the development of headaches in real time.

By understanding the science of brain freeze, scientists are discovering how headaches and migraines can be controlled using specific vasoconstriction or dilation medicines.

The next time brain freeze ruins the enjoyment of your favorite ice cream treat, just remember that it's simply your body and brain trying to protect you. You might consider thanking it in return by eating a vegetable every once in a while!

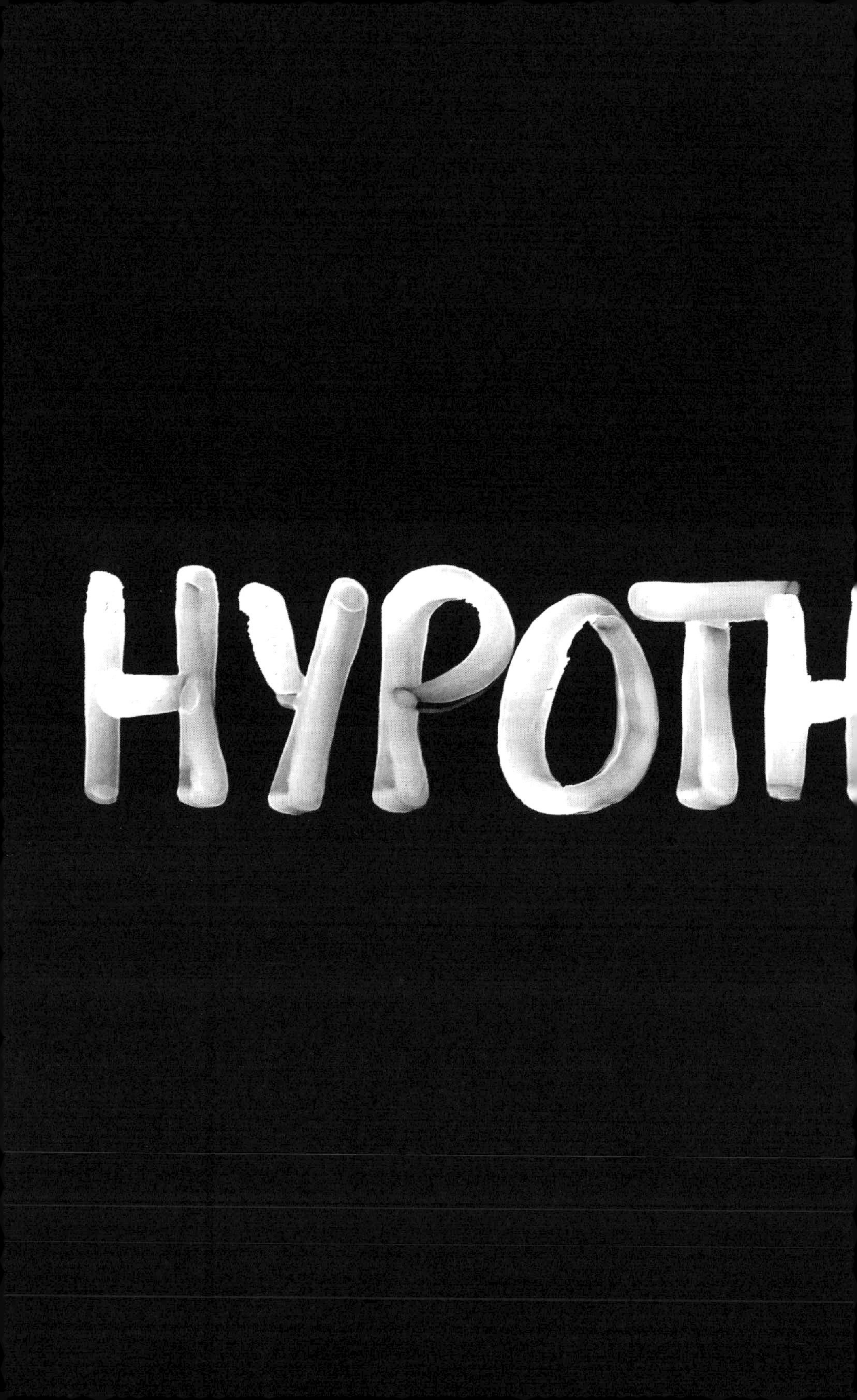
HYPOTH

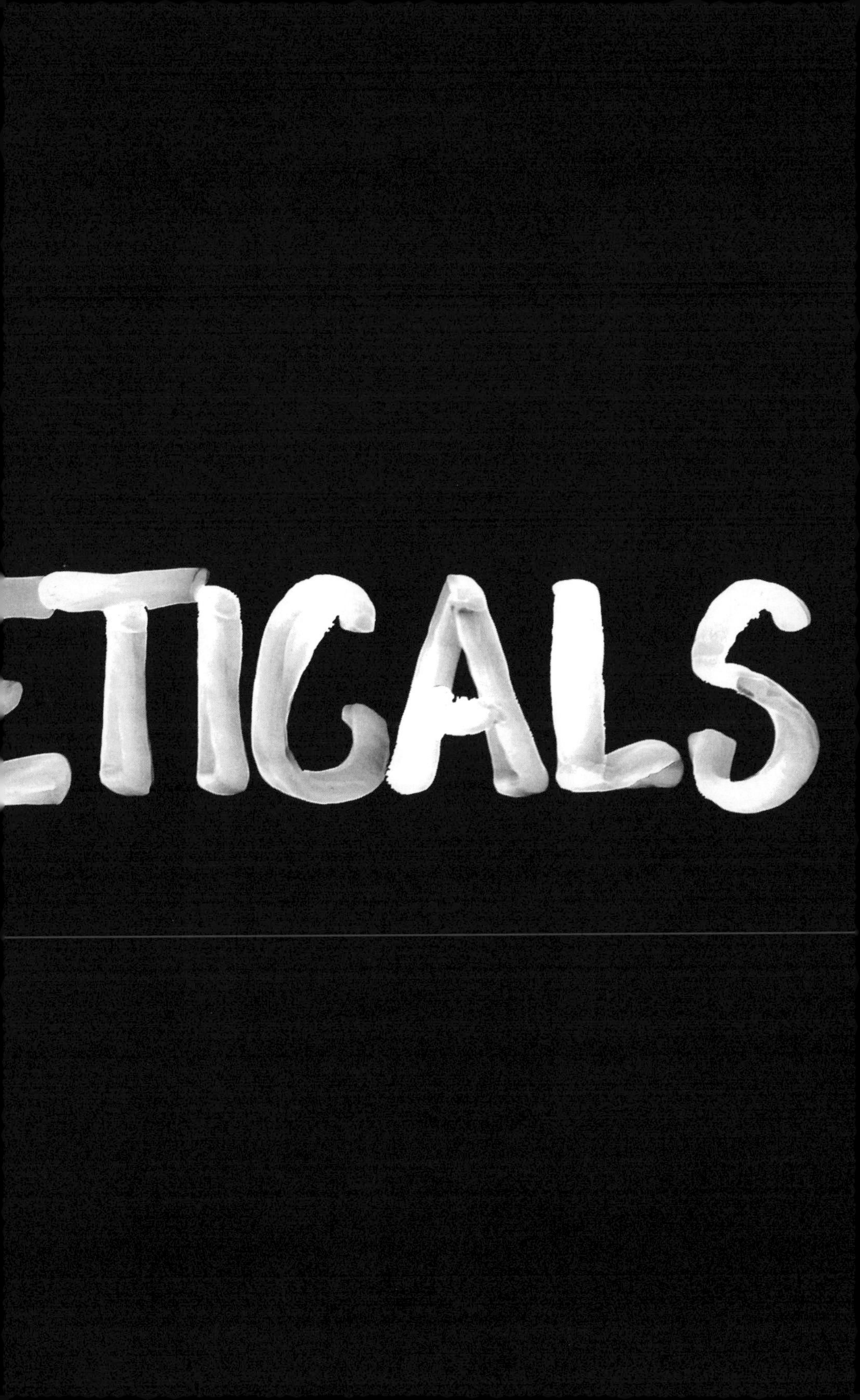
ETICALS

COULD A ZOMBIE APOCALYPSE HAPPEN?

History is full of stories of demons and other unworldly creatures wreaking havoc on society. But what does science have to say about zombies? Are these ferocious freaks plausible or scientifically accurate in any way? Should we be preparing for a zombie apocalypse?

Well, it depends on what type of zombie we're talking about.
Dying and coming back from the dead? Not so likely.
But a virus that induces the zombielike qualities of rage and
the incessant need to eat human flesh? Hypothetically,
yes. If, say, we found a virus that targeted the specific areas
in the brain that would cause these symptoms while
leaving other areas intact.

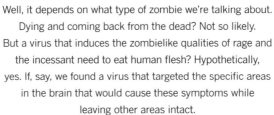

Viruses can enter the body and
affect cells in many ways, but we're
going to focus on the neurons.
Neurons are some of the longest
cells and they can transport
molecules and proteins throughout
the entire body.

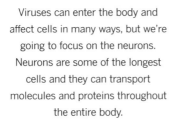

Many viruses, such as rabies, take advantage of a process called retrograde axonal transport, in which the virus travels from its entry point to where it wants to be. The rabies virus enters the body via a bite and slowly makes its way to your brain or central nervous system.

The farther it starts physically from your brain, the longer it takes to get there—sometimes years! But once it's there, it's generally too late. Many viruses follow a similar process, with the only difference being which neurons they're specific to. So the zombie virus would have to use specific neurons to affect particular parts of the brain to induce a zombie state without damaging the entire brain. Sound unlikely?

Well, it just so happens that the olfactory nerve, which is in your nose and used for scent, leads directly to some areas of the brain that could cause zombified effects and would leave other areas of the brain intact.

Specifically, it leads to the ventromedial hypothalamus, which tells you when you're full; the amygdala, which controls emotion and memory; and the frontal cortex, responsible for problem solving and long-term planning, morality, and inhibiting impulsive action.

Essentially, transmission of a virus through the olfactory nerve could create super-hungry, aggressive, brain-dead beings that can't recognize family and friends or control their actions other than to feed. Sounds kind of like a zombie!

So while the idea of a rotting or an invincible zombie may be unlikely, with the right virus nerve specificity, a form of zombies could be possible.

WILL YOU BE READY?

WHAT IF YOU STOPPED GOING OUTSIDE?

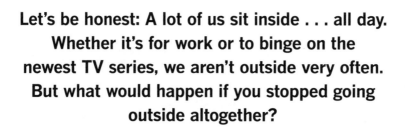

Let's be honest: A lot of us sit inside . . . all day. Whether it's for work or to binge on the newest TV series, we aren't outside very often. But what would happen if you stopped going outside altogether?

92,955,807 miles away, light is being ejected from the sun, shooting across the solar system, through our atmosphere, and if the timing is right, landing directly on your skin. Feels good, doesn't it? And it's this sunlight that begins an amazing chain reaction that helps to sustain your life: the "sunshine vitamin"—vitamin D, that is.

Interestingly, some of the cholesterol you consume is altered and stored in your skin. And when that well-traveled ultraviolet B sunlight hits your skin, it modifies this cholesterol. The new molecule travels through your bloodstream to your liver, where it is altered again, and then to the kidneys, where it is biologically "activated."

THE SUNSHINE VITAMIN

This activated vitamin D works to absorb calcium from your food, ultimately leading to bone growth and strength.

In a way, your skin eats the sun and the sun grows your bones! Perhaps we aren't so different from plants.

Without vitamin D, your body would not only suffer decreased mineralization of your bones, leading to diseases like osteoporosis, but also experience a decrease in immune function. Furthermore, there is evidence to suggest that vitamin D helps prevent cancer, heart disease, and depression. Which may explain why people in colder climates with less daylight often experience the winter blues.

But sunlight and vitamin D aren't the only reasons to venture outside. Many studies have shown that being in nature can have a mental and physical effect on the body. While using sophisticated brain-imaging techniques, researchers saw that when people looked at scenes in nature, their brains exhibited more activity in the regions associated with stability, empathy, and love.

On the other hand, viewing man-made environments produced activity in the regions associated with fear and stress.

On top of it all, if you are often inside, chances are you're sitting down. Admit it, you've been lounging, reading this book, and/or watching YouTube for hours.

Which may seem harmless, but studies have shown major adverse health effects linked to sitting time, such as an increase in type 2 diabetes and cardiovascular disease. Furthermore, a study of more than two hundred thousand people found a strong correlation between mortality and increased sitting time—in other words, the longer you sit, the more likely you are to die prematurely. The scariest part is that this is regardless of your physical activity.

So stand up, get outside, and live a little . . . longer than you would have. After finishing this book and watching all our AsapSCIENCE YouTube videos, that is.

SENSORY

PERCEPTION

WHY DO WE GET
PINS AND NEEDLES?

Whether you are watching YouTube for hours with
your legs crossed, or sleeping on your arm the
"wrong way," you have no doubt experienced pins
and needles—sometimes referred to as your
[insert body part here] falling asleep. In the
moment, nothing seems worse. And though this
pain seems easy to alleviate and forget, what
is happening in your body to cause such an
irritating sensation?

Your nervous system is composed of the brain, the spinal cord, and the nerves. The brain and spinal cord comprise the central nervous system. This is the control center of your body, in charge of everything you do consciously along with everything your body does automatically, like breathing.

The brain and spinal cord are made of complex nerve bundles you can think of as busy information highways in your body. The rest of the nervous system is made up of the nerves that attach your brain and spinal cord to your extremities. When you are moving your pinkie finger or toe, the nerves carry the information to the peripherals of your body. Think of your nerves as smaller roads leading off of the brain and spinal cord highways.

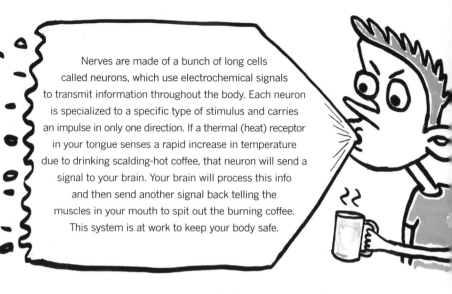

Nerves are made of a bunch of long cells called neurons, which use electrochemical signals to transmit information throughout the body. Each neuron is specialized to a specific type of stimulus and carries an impulse in only one direction. If a thermal (heat) receptor in your tongue senses a rapid increase in temperature due to drinking scalding-hot coffee, that neuron will send a signal to your brain. Your brain will process this info and then send another signal back telling the muscles in your mouth to spit out the burning coffee. This system is at work to keep your body safe.

When a part of your body "falls asleep," it means there is a disturbance or traffic jam in the pathway of your nervous system. When you cross your legs or sleep on your arm, you are cutting off the blood circulation to specific nerves.

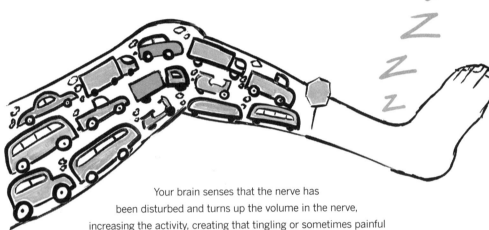

Your brain senses that the nerve has been disturbed and turns up the volume in the nerve, increasing the activity, creating that tingling or sometimes painful sensation. This can be relieved by taking away the pressure and allowing the blood to flow back to the nerve. The feeling you get when you hit your "funny bone" happens in a similar way; when you bump the ulnar nerve in your elbow, the disturbed nerve reacts with a painful sensation because the nerve has been hit directly.

If you experience pins and needles frequently, you may have neuronal problems or damaged neurons indicating a more serious issue.

Less frequent pins and needles are a natural sensation created by your body to alarm the nervous system when it thinks something is wrong. So go ahead and wake those neurons up!

WHY DO WE ITCH?

Bugs, scabs, rashes, and even dust can all create the insatiable desire to scratch! But even conditions such as depression or obsessive-compulsive disorder can create this same sensation. Scratching your own body is a bizarre thing, so why exactly do we itch?

The average human body is covered in about
eighteen to twenty-two square feet of skin.

4.5'

4.5'

Once thought simply a barrier
for the body, we now know
skin can heat you up, cool you
down, transform sunlight
into vitamin D, and relay
sense of touch to your brain.

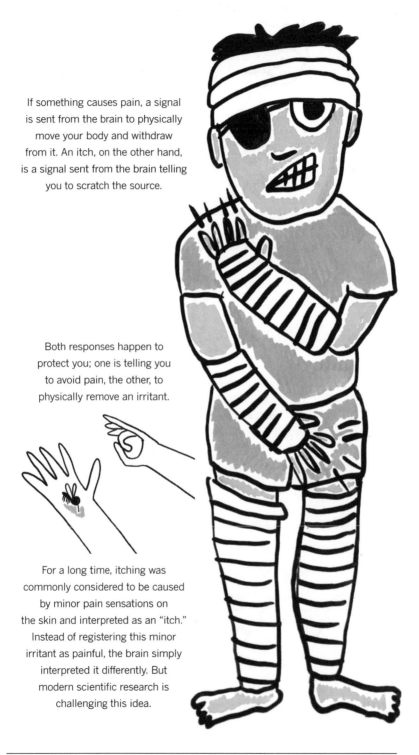

If something causes pain, a signal is sent from the brain to physically move your body and withdraw from it. An itch, on the other hand, is a signal sent from the brain telling you to scratch the source.

Both responses happen to protect you; one is telling you to avoid pain, the other, to physically remove an irritant.

For a long time, itching was commonly considered to be caused by minor pain sensations on the skin and interpreted as an "itch." Instead of registering this minor irritant as painful, the brain simply interpreted it differently. But modern scientific research is challenging this idea.

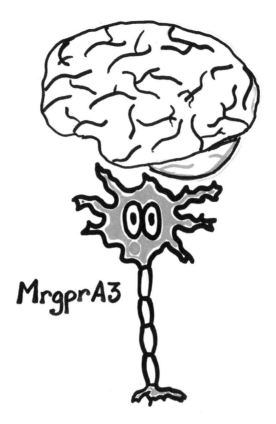

Scientists have discovered that a specific
type of neuron (called the MrgprA3 neuron) is
responsible for detecting an itch.

When these neurons are destroyed in
mice, they no longer scratch themselves
if exposed to itchy substances.

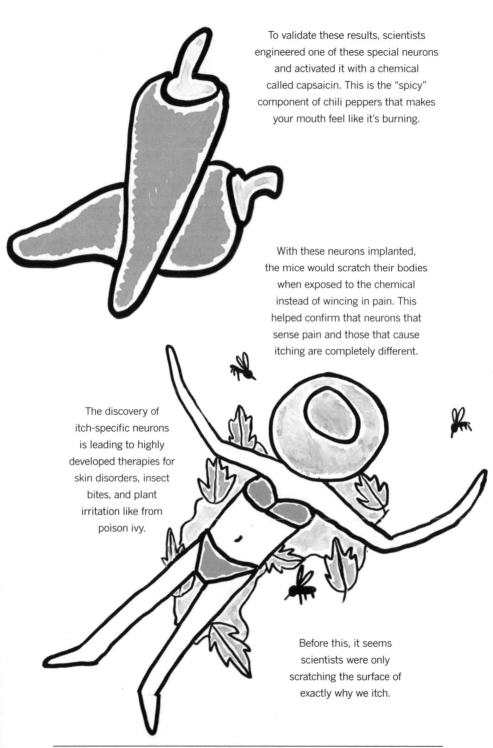

To validate these results, scientists engineered one of these special neurons and activated it with a chemical called capsaicin. This is the "spicy" component of chili peppers that makes your mouth feel like it's burning.

With these neurons implanted, the mice would scratch their bodies when exposed to the chemical instead of wincing in pain. This helped confirm that neurons that sense pain and those that cause itching are completely different.

The discovery of itch-specific neurons is leading to highly developed therapies for skin disorders, insect bites, and plant irritation like from poison ivy.

Before this, it seems scientists were only scratching the surface of exactly why we itch.

SELFIE SCIENCE:
WHY DO WE HATE
PHOTOS OF OURSELVES?

You're ready for a big night out and take one last
look in the mirror before you go. Wow, you're
looking great tonight! With all the confidence in
the world, you strut your stuff and make sure
to take tons of photos through the night—selfies
galore. But wait . . . you hate how you look
in photos and feel completely unphotogenic. So
how come you never really feel that way looking
in the mirror? Why is it that so many people
hate seeing photos of themselves but usually have
no problems with their image in a mirror?

MERE EXPOSURE EFFECT

There are a few factors at play with this phenomenon, but the mere exposure effect may be the biggest. Simply put, we all have a tendency to prefer familiar things.

After repeated exposure to *anything*, you will psychologically prefer it over a version that you have seen less often. As crazy as it may seem, this has been tested with words, paintings, sounds, pictures, and even geometric figures.

You look familiar!

This phenomenon has even been observed in other species: From monkeys to chickens, many organisms have been documented to react fondly to a repeated stimulus!

And it just so happens that the version of yourself that you see most often is your mirror or reflected image. Every time you stare in a mirror (or window reflection), the view is flipped from left to right. If your reflection in the mirror were actually somebody you were looking at, his hair would swoop the opposite way, when you raised your right hand he would raise his left, and even that annoying pimple you have would be on the opposite side of his face.

But a photograph is *not* your reflected image. In fact, your photo image is the way everybody else sees you on a daily basis. Your brain, however, isn't familiar with this view of you and might interpret it as "off."

In studies of this phenomenon, participants preferred facial photographs of their mirror image to their photo image, without knowing which was which. Their friends, however, rated their photo image as preferable. So if your acquaintances and loved ones suddenly saw your mirrored self, they would find it slightly off.

Add this to the fact that you can adjust your position, hair, and smile in front of the mirror, and that photographs can capture angles that you can't see in a reflection, and you've got a good case against your photo image. But just know that there is nothing to worry about—the photo version of you looks great to everybody else!

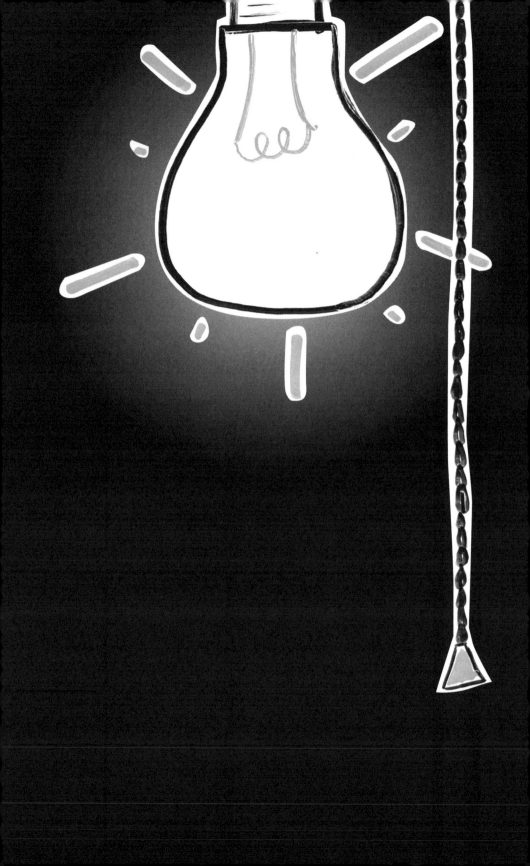

WHERE DOES ALL THE LIGHT GO WHEN YOU TURN OFF THE LIGHTS?

It seems simple enough—you flick a switch, and in the blink of an eye, the room has gone completely dark. But where exactly did all of the light go? It was in the room one second ago, so what actually makes the room dark?

According to the conservation of mass and energy (a scientific law that some really smart people discovered), light can't simply "disappear." In fact, no matter can vanish. Instead, it only changes its form, composition, color, or other property, ultimately ending up as something else. So how does this apply to light?

Light is a form of energy that consists of tiny particles called photons. When you flip on a lightbulb, the filament is heated to the point at which it emits these photons.

They shoot around the room in all different directions at incredible speeds—the speed of light, to be exact, which is 186,000 miles per second.

On the other hand, the walls of your room—and everything else, for that matter—are made up of a structure of atoms. Picture atoms like marbles all organized beside each other. Whenever a photon hits one of these atoms, it transfers its energy.

And since this quantity of energy needs to be conserved (remember, nothing can just disappear), the atom gets a jolt of energy, which causes it to vibrate. This vibrating atom then knocks into its neighboring atom, causing it to vibrate as well, which causes a ripple effect of energy being transferred from one atom to the next.

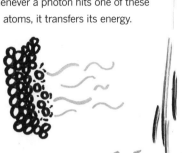

It's kind of like dropping a rock into a pool of water. The energy from the rock spreads to the atoms in the water and then ripples outward to more and more water molecules. This heats up the wall ever so slightly, similar to how sunlight can heat something. In fact, you can put your hand near a lightbulb and feel the heat. So when the light is on, the photons in the room are hitting the wall and absorbed as thermal energy.

What does this have to do with the darkness? Well, when you turn the light off, the remaining photons in the room hit the walls and other objects and are absorbed as thermal energy. And because the speed of light is so incredibly fast, it all happens in an instant. It's that simple! And you are left in darkness and fear, stumbling over the mess that is your life.

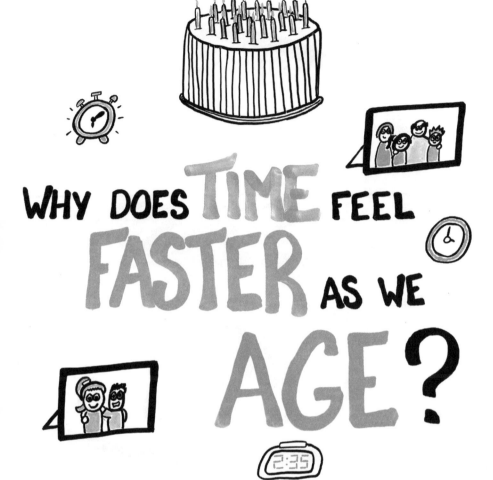

WHY DOES TIME FEEL FASTER AS WE AGE?

We all get older. But don't you sometimes get the feeling that time is passing much more quickly now than it did when you were younger? Long gone are the summer holidays that felt like they lasted forever. Instead, birthdays seem to pop up sooner than ever, and events from years ago seem like they happened yesterday. So why does our perception of time change so drastically with age? And is there anything we can do about it?

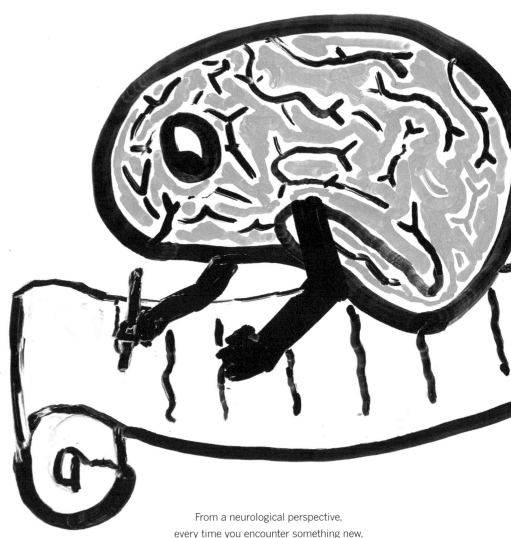

From a neurological perspective,
every time you encounter something new,
your brain tries to record as much
information as possible.

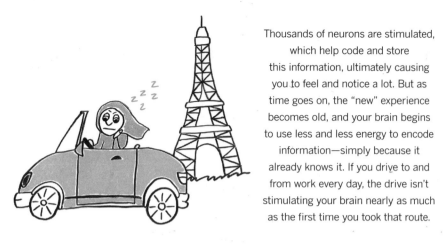

Thousands of neurons are stimulated, which help code and store this information, ultimately causing you to feel and notice a lot. But as time goes on, the "new" experience becomes old, and your brain begins to use less and less energy to encode information—simply because it already knows it. If you drive to and from work every day, the drive isn't stimulating your brain nearly as much as the first time you took that route.

The key is novelty: It's the first time you experience something that the brain remembers most, which ultimately creates greater detail in the brain.

Of course, we experience most of our "firsts" during the earlier portion of life, which contributes to the overwhelming feeling that much more happened when we were young. Whether it was your first kiss,

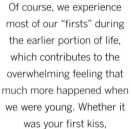

first bike ride,

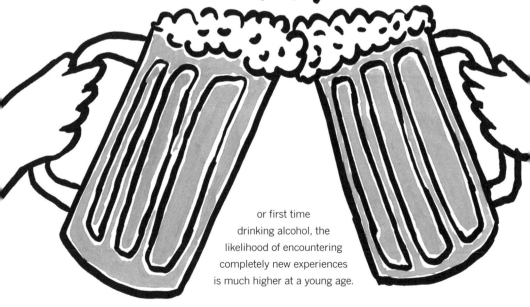

or first time drinking alcohol, the likelihood of encountering completely new experiences is much higher at a young age.

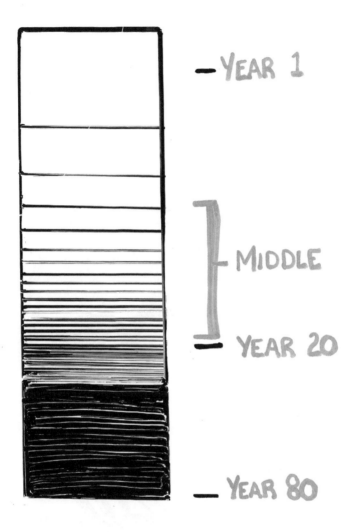

YEAR 1

MIDDLE

YEAR 20

YEAR 80

Add to that the decreasing proportion of time that one year represents in your life, and the feeling is compounded. Think about it this way: When you turn one year old, that one year in your life represents 100 percent of the time you have been alive. Everything you've ever experienced happened in a single year. Jump ahead to fifty years old, and one year in your life now represents only 2 percent of your life. Proportionately, a year represents a smaller fraction of your life as a whole, and as a result it feels as if it went by faster. If we built a chart using this theory for each successive year, we'd see the first year is 100 percent of your life, the second year is half of your life, the third is one-third, the fourth one-quarter, etc. Here, we see that when you turn eighty years old, you will perceive the "middle" of your life to be just before your twenties—from your brain's time perspective, that is. Of course, this is just one of many theories.

But all hope is not lost. If you continue finding novel experiences throughout old age—things that stimulate new parts of your brain—time may feel slower again.

By learning a new language, traveling,

or doing activities you've never done,
your brain can escape the monotony that
life sometimes brings.

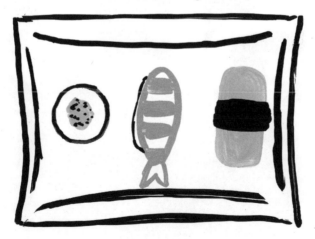

By adding more new experiences, you can amplify the
feeling that time is moving as slowly as it did back when
summer seemed to last forever.

AND OTHER
AMOROUS
PURSUITS

THE SCIENCE OF
SEXY

Hot? Or not? There are so many ways we can
judge each other, none more quickly than
how someone looks. Though we may not want to
admit it, seeing someone attractive can be
exhilarating and make your heart pound! But
can beauty or sexiness be objectively measured?
Or is one person's trash another person's
treasure? What is the science of sexy?

Scientifically speaking, humans are on Earth to survive long enough to reproduce and then die, all in hopes of perpetuating the human race. Like many species, our genes drive us to want to reproduce with sexual partners that are top-notch biologically. For early humans, being able to identify a fit and healthy mate was key to the survival of our species.

Studies have shown that men are attracted to women with long legs, small waists, and large hips. Long legs represent increased fitness, while large hips suggest better childbearing capabilities, healthier offspring, and potentially heavier babies.

Women find men with certain facial scars to be more sexually appealing. The scars are theorized to imply high testosterone levels, strength, and masculinity. Our ancestors may have seen these factors as beneficial to fight and defend a family or pass on to an offspring. Perhaps a link to why some women prefer a "bad boy."

On a more personal note, studies have shown that selflessness and loyalty are preferred traits in men because they are thought to indicate a willingness to stick around and be a long-term protective parent, something that was essential to the survival of our early ancestors' offspring. So maybe nice guys don't always finish last!

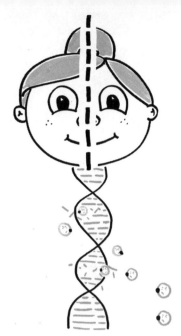

Universal beauty may be defined as how symmetrical our bodies are. Many studies have shown that when a person's features match up evenly on both sides of their body, they are perceived as more attractive. When someone's features are not symmetrical, it can be an indication of developmental issues in the womb. The damage happening in the womb can be caused by free radicals, which are biologically damaging molecules affecting DNA and creating a variety of diseases, including cancer.

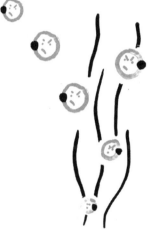

Smoking while pregnant, obesity, and birth complications can create an abundance of free radicals in the mother that decrease the symmetry of the baby. So a person's visible symmetry may be an indication of healthy development and strength against diseases, all factors attractive to a potential mate.

So it seems that our ancestors' quest for a biologically fit mate and healthy offspring may still be relevant to our ideas about what sexy is today.

THE SCIENCE OF
HEARTBREAK

The end of a relationship can be devastating.
Feeling tired, lonely, and even depressed
are common emotional reactions to heartbreak.
But is a "broken heart" simply an abstract
concept, or are there real physical effects on the
body and brain?

When you endure physical pain, such as a cut or injury, the anterior cingulate cortex is stimulated. Surprisingly, it's this same region of the brain that's activated when you feel excluded or experience the loss of a social relationship. Perhaps physical pain and emotional pain aren't as different as we once thought.

Think about the ways we describe lost love:
"He ripped my heart out,"

RIPPED MY HEART OUT!

SLAP IN THE FACE!

"My heart is broken,"
"It was a slap in the face,"

"I'm emotionally scarred for life."

EMOTIONALLY SCARRED

The use of physical description paints a clear relationship, at least in language, between emotional and physical pain. In fact, studies have shown that we would rather be physically hurt than feel social exclusion. But *why* do these two different experiences elicit the same feeling in our bodies?

It's clear that our bodies use physical pain to diminish the risk of imminent danger. From an evolutionary perspective, anything that increases our overall survival and fitness as a species is likely to persist. The rise of relationships and social bonds between lovers and friends became an important part of survival for many species.

You look out for me, and I'll look out for you! And just like your desire to not be burned by hot coffee again, animals desire not to be socially alone. You're more likely to survive and reproduce if you're not alone.

This can be seen directly in studies of primates, who, when separated from loved ones, experience an increase in the hormone cortisol and a decrease in the hormone norepinephrine, which reflects a major stress response.

Ultimately, this contributes to depression, anxiety, and loud crying.

For humans, a breakup, loss of a loved one, or isolation can trigger a similar reaction, creating the perception of physical pain.

So how can we alleviate this pain? After all, bandages and creams are meant for physical wounds; there's nothing you can get at the pharmacy that can mend a broken heart. Studies have shown that high levels of social support are related to lower levels of pain, whereas socially alienated individuals show poor adjustment and experience higher levels of pain.

So if you're feeling brokenhearted, surround yourself with friends and
family. And if someone you know is suffering emotionally,
be there for social support—because scientifically, we humans all
just want to fit in somewhere.

THE SCIENCE OF
LOVE

From philosophers and historians to poets and scientists, love has captured our imagination and curiosity for centuries. Many of us have experienced the rush of falling in love for the first time, or the deep feelings of love for children, family, or friends. But what is love, from a biological perspective?

No doubt it's intertwined with the evolutionary survival of our species. After all, you come from an unbroken line of organisms reproducing, from the very first microbe that split in two, to your ancestors who have all successfully mated since the dawn of time.

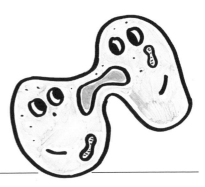

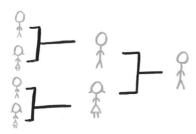

Unfortunately, if you fail to have children, this perfect streak comes to a halt. While we're driven to reproduce, we are also driven to make sure our offspring survive.

Though we often associate love with the heart, the true magic can be seen inside the brain. It may not be entirely surprising to find out that the brain of somebody in love looks awfully similar to one on cocaine.

Cocaine acts on the pleasure centers in the brain by lowering the threshold at which they fire. This means that you feel really good, a lot easier. We see the same thing in the brains of those in love. But it's not just the cocaine or the love that makes you feel good; it's the fact that anything you experience will now more easily set off pleasure centers and make you feel great. Because of this, you not only fall in love with the person but at the same time begin to build a romanticized view of the world around you. Interestingly, nearby pain and aversion centers begin to fire less—so you become less bothered by things. Simply put, we love being in love!

So what chemicals are at work to make all of this happen? Both during orgasm and by simply looking at photos of a loved one there is a surge of dopamine and norepinephrine from the ventral tegmental area of the brain.

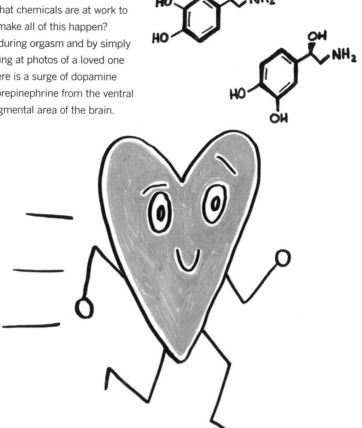

This not only triggers sexual arousal and your racing heart but also gives you the motivation, craving, and desire to be with a person more and more. You see, romantic love is not simply an emotion—it's a human instinct rooted in the evolution of our species that led to greater survival. And this drive brings about intense energy, focused attention, and elation. The pleasure centers are part of the brain's reward system—the mesolimbic dopamine system. If you stimulate this region while learning, learning becomes much easier, because it's pleasurable and perceived as a reward.

OXYTOCIN

We also see a surge in the neuromodulator oxytocin released from nucleus accumbens—sometimes called the commitment neuromodulator because in mammals it helps to reinforce bonding or attachment. When prairie voles are injected with either oxytocin or the hormone vasopressin, they will instantly find a mate to pair-bond with.

Finally, studies have shown that people in love have low levels of serotonin—which is similar to people with obsessive-compulsive disorder. This is the likely cause of our obsession and infatuation during early love.

Amazingly, these areas associated
with intense romantic love can
remain active for decades after you
first fall in love. And while there
are many other physiological and
psychological components that
add to this mix, the truth is, science
still knows very little about exactly
why or how love works. And
yet somehow we all seem to know
it when we feel it.

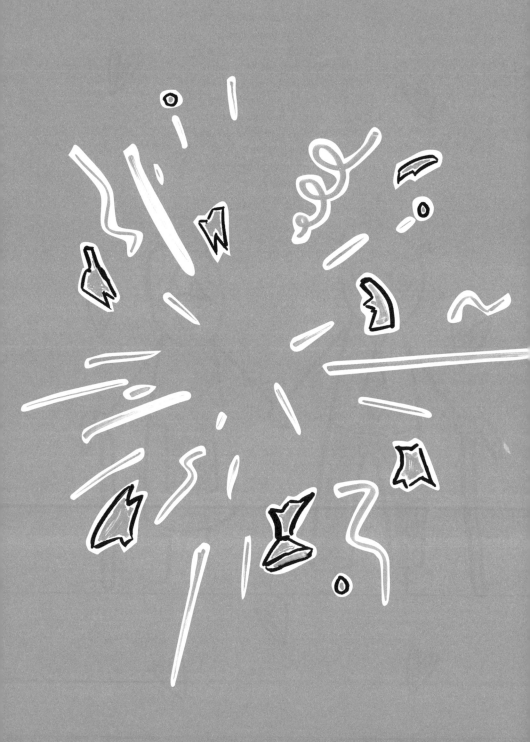

THE SCIENCE OF
ORGASMS

The human body is a wonder, but perhaps the
most curious and quintessential aspect of
the human experience is the orgasm. So why does
it feel so good? Read on!

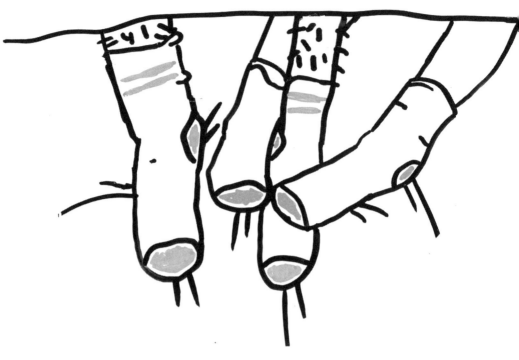

The body's sexual response is typically broken down into four stages: excitement, plateau of arousal, orgasm, and resolution. Following arousal, the brain stimulates blood flow to your genitals, your heartbeat and breathing increase, and your central nervous system is fully engaged, sending signals of enjoyment to your brain's reward system. Thousands of nerve endings constantly relay pleasure signals to your brain, resulting in an orgasm.

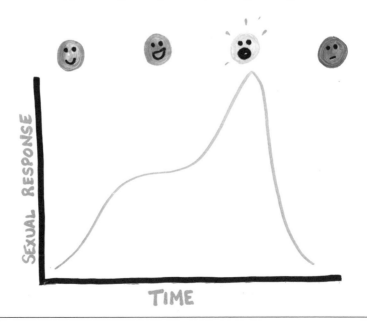

3 - 10 SEC
PLEASURE

For men, the orgasm includes rapid contractions of the anal sphincter, the prostate, and the muscles of the penis. In conjunction with ejaculation, the release of sperm seminal fluid, the whole process for men involves around three to ten seconds of intense pleasure.

This is followed by a refractory period from minutes to hours, in which another orgasm cannot be achieved.

20+ SEC
PLEASURE

Women, on the other hand, do not experience a refractory period, allowing them to experience multiple, consecutive orgasms. On average, these last around twenty seconds, though sometimes much longer, and consist of rhythmic contractions of the uterus, vagina, anus, and pelvic muscles.

But it's the brain that takes control, or rather, lack thereof, during orgasm. Using functional MRI scans, scientists are able to see activity in more than thirty discrete regions of your brain.

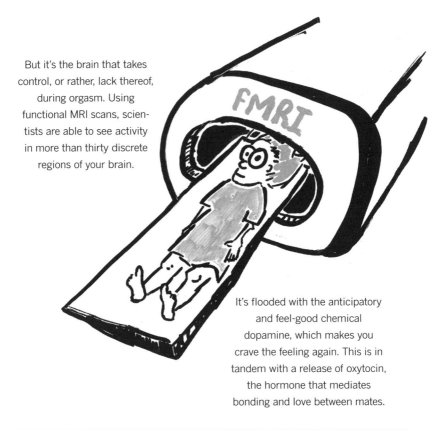

It's flooded with the anticipatory and feel-good chemical dopamine, which makes you crave the feeling again. This is in tandem with a release of oxytocin, the hormone that mediates bonding and love between mates.

PET scans show, surprisingly, that brain activity during an orgasm is the same in men and women. In both genders, the lateral orbitofrontal cortex, which controls self-evaluation, reason, and control, is turned off. This mechanism shuts down fear and anxiety, which is an essential aspect of having an orgasm.

The relaxation of the amygdala and hippocampus in women further reduces emotions, producing a trancelike state; in men, it dampens aggressiveness.

Many areas of a woman's brain are shut down completely during an orgasm. These effects are less striking in men, likely because of the shorter duration and subsequent difficulty in measuring with a brain scan. In women, an area called the periaqueductal gray is activated, stimulating the fight-or-flight response, while the cortex, which is associated with pain, lights up, suggesting that there is a connection between pain and pleasure.

Following the climax and muscle contraction, the body experiences deep relaxation and the heart rate slows to a resting pace.

Who knew science could be so sexy!

WILL DANCING GET YOU LAID?

Throughout history and across all cultures, whether it's a tribal ritual, the two-step, or twerking, we all have a personal way of expressing ourselves on the dance floor . . . or at home in front of the mirror. And while some of us are seemingly a lot better than others, the question remains: Are these expressions actually individual or part of an embedded human code? Why are we so drawn to dance, and is it really equivalent to a human mating call? More important, does dancing increase your potential to . . . get some?

Well, from a viewer's perspective, dance can be pleasurable to watch for many of the same reasons music is enjoyable to listen to. But it was the science legend Charles Darwin who first suggested that dance could be a part of mating rituals and entangled with the evolution and survival of our species. But if this is true, the ability to dance must somehow relate to the ability to survive and reproduce. So . . . does it?

It turns out that early humans who were coordinated, symmetrical, and rhythmic were seen as more viable to potential mates, because these traits were often linked to survival— the ability to run faster, defend against prey, and, ultimately, survive.

And dancing just happens to be an outward manifestation and display of many of these qualities, a quick external way of documenting and proving your fitness in order to attract a mate.

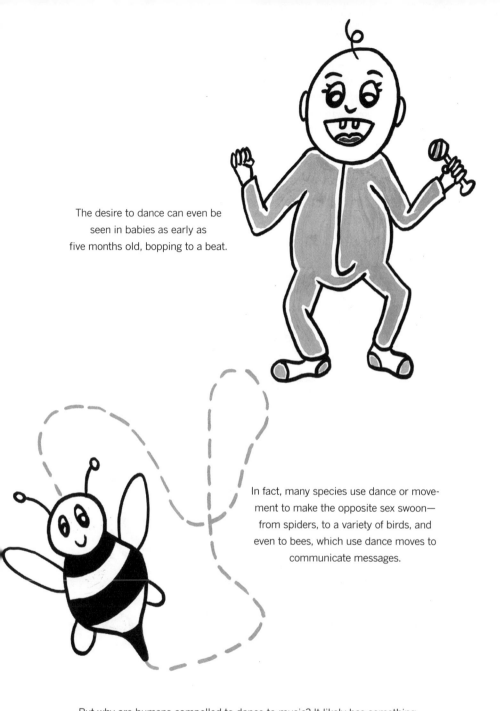

The desire to dance can even be seen in babies as early as five months old, bopping to a beat.

In fact, many species use dance or movement to make the opposite sex swoon—from spiders, to a variety of birds, and even to bees, which use dance moves to communicate messages.

But why are humans compelled to dance to music? It likely has something to do with the fact that many of the brain's reward systems are directly connected to the motor areas. For example, music stimulates the cerebellum at the base of the brain, which is involved in the timing and coordination of movement.

In a contemporary sense, dance is known to show confidence and fearlessness . . . and sometimes intoxication. But most important, movements associated with good dancing are indicative of good health and reproductive potential. And in many cases, the human courtship ritual in our "natural habitat" of a nightclub mirrors that of animals in the wild. This gives "being an animal on the dance floor" a new meaning, except our liquid courage doesn't come from the local water hole.

Many men feel as though they can't dance or that dancing isn't for them. But, ironically, in most animals, the males must impress the female. So what, then, is "good" dancing? Scientists have actually studied which male dance movements are most attractive to women. Interestingly, they found that twisting, bending, and fluid movements of the core body region (the torso, neck, and head) garnered the most positive attention from women. Showing variability, flexibility, and creativity was perceived as good dancing. On the other hand, repetitive, rigid, and twitchy movements weren't particularly attractive (sometimes called "dad dancing"). And truthfully, your dad probably dances a little more slowly and rigidly than his youthful counterpart, because, simply put, he isn't as youthful or fit anymore.

So whether you're in a club, a bar, or your own living room, engage and embrace your animal instincts! And if you're asked where you got those moves, just say "millions of years of evolution have led to this dance, baby!"

Who knew science could be so groovy?

GETTING TO THE
BOTTOM
OF
BAD
BEHAVIOUR

@&#?

SH✳T F#CK

THE SCIENCE OF

SWEARING

A$$

Why the #*&! do we swear? We all let one slip occasionally—or maybe you express your expletives loud and proud. But does swearing serve any biological purpose, or is it simply a cultural taboo?

While swear words are generally denounced as crude and potentially harmful, science says swearing may actually be good for you . . . under certain circumstances!

It's no surprise that people often swear when they are physically hurt. In fact, it's such a common response to pain that researchers began looking into the underlying reasons for these sudden outbursts. What they found was that swearing may actually play a role in pain modulation.

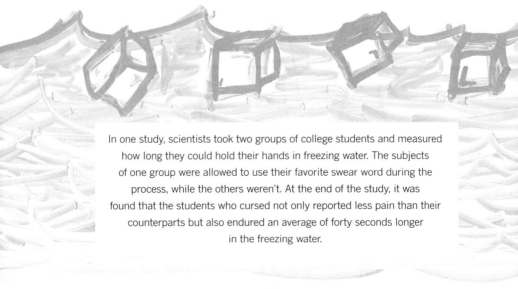

In one study, scientists took two groups of college students and measured how long they could hold their hands in freezing water. The subjects of one group were allowed to use their favorite swear word during the process, while the others weren't. At the end of the study, it was found that the students who cursed not only reported less pain than their counterparts but also endured an average of forty seconds longer in the freezing water.

Researchers speculate that swearing may tap into the brain circuitry linked to emotion. While normal, everyday language uses the outer left hemisphere of the brain, expletives may stimulate an area called the amygdala, in the right hemisphere.

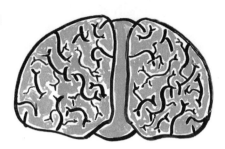

This happens to be where our fight-or-flight response is triggered, which in turn suppresses pain sensation. After all, in times of stress, the last thing your body wants to worry about is pain!

This defensive reflex is stimulated in many animals when they are scared, confined, or hurt, and is often accompanied by an angry vocalization in order to startle a potential predator. Kind of like when you accidentally step on your cat's tail, and it sends out a big *YELP!*

But there's a catch: The more we swear, the less meaningful and effective the words become. The emotional response is turned down. So while swearing may have evolved to save our lives, don't #&*@ing overuse it!

THE SCIENCE OF LYING

White lies, exaggerations, boldface lies, half-truths, lies by omission, bluffs—there are multiple ways to intentionally deceive someone. Although most of us lie daily, our lies differ in complexity, and the intentions of fibs vary, as do the stakes of getting caught. But is it possible to detect deception with 100 percent certainty? Do our bodies all respond in a similar biological fashion when we stray from the truth?

Telling a lie can be a stressful experience, especially when the consequences of getting caught are high. Your body actually reacts to the emotional stress and anxiety of telling a lie in the same way it does when you experience any environmental stressor or threat: Your fight-or-flight reflex is triggered.

FIGHT OR FLIGHT

Lying activates a defense mechanism in your nervous system because a threat is perceived, releasing neurotransmitters such as acetylcholine, adrenaline, and epinephrine, which act as chemical messengers, bringing certain areas of your body to action. You feel a surge of energy, and a range of changes take place:

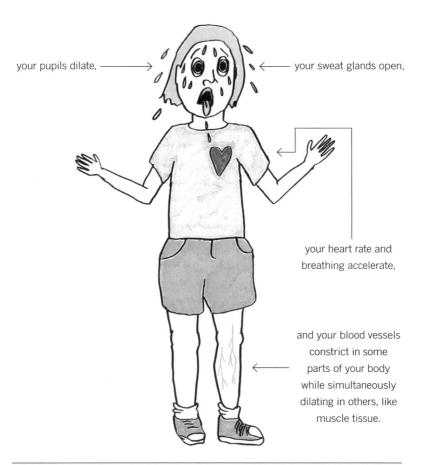

your pupils dilate, ⟶

⟵ your sweat glands open,

your heart rate and breathing accelerate,

and your blood vessels constrict in some parts of your body while simultaneously dilating in others, like muscle tissue.

For many years now, the polygraph has been the
instrument of choice to pick up on these changes
and detect lies. However, the major flaw in
these machines is that not all people experience
anxiety from lying in the same way.

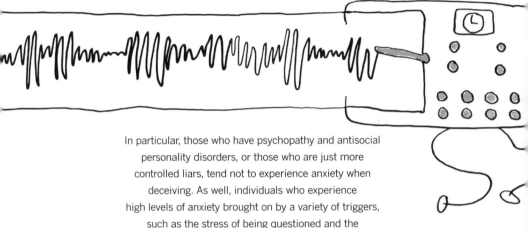

In particular, those who have psychopathy and antisocial
personality disorders, or those who are just more
controlled liars, tend not to experience anxiety when
deceiving. As well, individuals who experience
high levels of anxiety brought on by a variety of triggers,
such as the stress of being questioned and the
fear of being disbelieved, can create inaccurate spikes
in the polygraph.

As an alternative, functional MRI scans are being considered as a
tool to examine the main organ responsible for formulating a
lie in the first place: the brain. Typically used for diagnosing neurological
disorders and mapping the brain, fMRI scans identify brain activity
by calculating the changes in blood oxygen levels across brain tissue.

When neurons in a region of the brain are activated, blood flows
to the region, and oxygen levels increase. An fMRI makes
a series of scans that depict those changes in the flow, essentially
reading your mind.

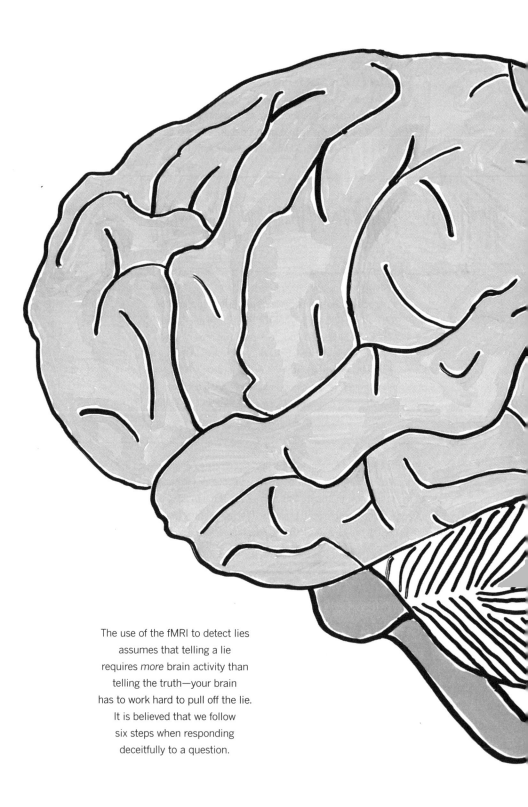

The use of the fMRI to detect lies
assumes that telling a lie
requires *more* brain activity than
telling the truth—your brain
has to work hard to pull off the lie.
It is believed that we follow
six steps when responding
deceitfully to a question.

Initially there must be **(1)** perception and **(2)** comprehension of the question being asked, followed by **(3)** memory recollection of information pertaining to the question. Next, **(4)** judgment, planning, and decision making come into play, as we must weigh the risks against the benefits and decide between responding truthfully or not. Since it is believed that the brain automatically generates a truthful response, **(5)** active suppression of the truth or the response inhibition is required, followed by **(6)** delivery of the lie.

It has consistently been shown in studies using fMRI, that there is a higher level of brain activity during a lie. This is especially true in the regions that are responsible for control and decision making.

Unfortunately, there is no universal method for a layperson to detect a lie since we (thankfully) lack noses that grow upon lying. However, advances in technology have provided an array of methods to assist scientists in distinguishing truth tellers from liars.

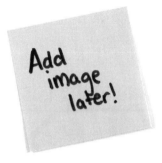

THE SCIENCE OF
PROCRASTINATION

Let's face it, you're likely reading in an effort to avoid some other task. You're procrastinating to learn how to stop procrastinating (it's called procrastination inception!). But the clock is ticking. So why can't you seem to stop?

Though the psychological causes are still debated, most of us tend to over- or underestimate the value of a reward based on its temporal proximity. This is often referred to as temporal discounting. For example, if we offered you $100 today, or $110 in a month, you most likely would take the $100 now. But what if, instead, we offered you $100 in a year, or $110 in a year and one month. Suddenly, you might say to yourself, "If I can wait a year, I can wait the extra month." But the time *and* value difference are the exact same in each example.

It turns out that human motivation is highly influenced by how imminent the reward is perceived to be. The further away the reward is, the more you discount its value. This is often referred to as present bias or hyperbolic discounting. So reading, or being on Facebook, reddit, Twitter, or YouTube today is more rewarding than a perfect score on your test, until temporal proximity increases the value of a good grade on your test . . . and you cram all night.

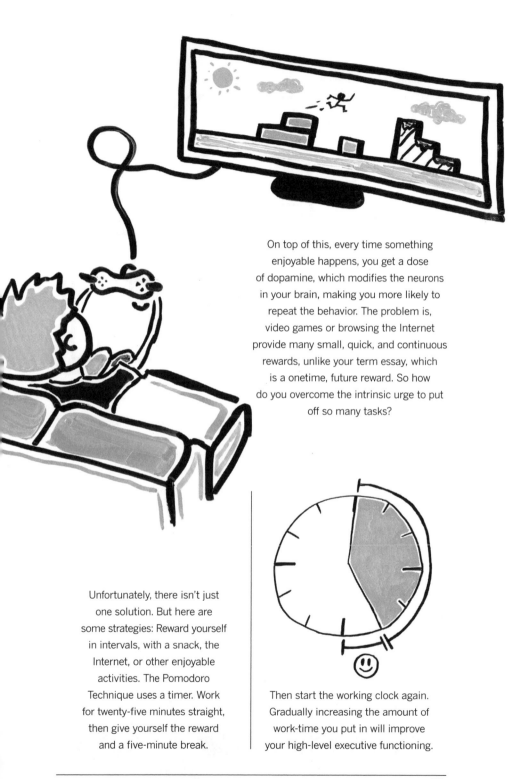

On top of this, every time something enjoyable happens, you get a dose of dopamine, which modifies the neurons in your brain, making you more likely to repeat the behavior. The problem is, video games or browsing the Internet provide many small, quick, and continuous rewards, unlike your term essay, which is a onetime, future reward. So how do you overcome the intrinsic urge to put off so many tasks?

Unfortunately, there isn't just one solution. But here are some strategies: Reward yourself in intervals, with a snack, the Internet, or other enjoyable activities. The Pomodoro Technique uses a timer. Work for twenty-five minutes straight, then give yourself the reward and a five-minute break.

Then start the working clock again. Gradually increasing the amount of work-time you put in will improve your high-level executive functioning.

Acknowledge your procrastination: Your future self will procrastinate. It's been shown that creating a self-imposed, costly deadline is an effective way to manage your working habits. Externally imposed deadlines are even more effective.

Also, enjoy the process of achieving something—instead of thinking, "Only twenty more minutes of torture," think, "I'm doing something great" or "I enjoy being productive." By the same token, make a list of the reasons you *want* to reach the goal. Reinforcing that you want to do it minimizes indecision. Often, procrastination is a symptom, not a cause, and the power of being properly motivated can take you far.

Finally, if you can, remove the temptations.
Turn off the Internet, uninstall your
favorite game, or work somewhere else.

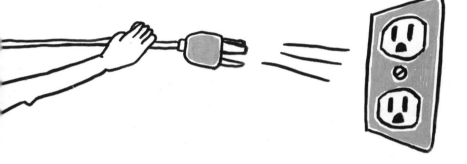

Putting obstacles in the way of your
procrastination tools can be a great trick
to keep you on track.

THE SCIENTIFIC
HANGOVER CURE

So you had a night out and woke up feeling . . . not so great. The solution? Stop drinking . . . until next weekend when it happens all over again. It's inevitable and you know it. There are many pills and myths that claim to do the job, but here are the actual scientifically proven tips to help handle your hangover.

I. BEFORE YOU START DRINKING

Eat: fatty foods and carbs. Sure you want to look slim on your night out, but the fact is that fatty foods slow the absorption of alcohol by your body and help curb stomach irritation, while carbs prevent low blood sugar and ease nausea, which means ixnay on the omitvay (no more vomiting!). This also gives your body more time to process the harmful by-products of alcohol—one of the main reasons for your hangover.

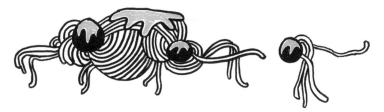

Next: water. Water is your best friend before, during, *and* after drinking. Alcohol is a diuretic, which means you'll be peeing out more fluid than you're taking in. And with less water in your body, your other organs start to steal it from the brain, causing it to literally shrink, which causes headaches. *Ouch!*

And while any pill that claims to fully cure hangovers is most likely a load of crock, the only pills that *may* help a bit are multivitamins. Because of the diuretic effect of alcohol, you'll be peeing out these essential vitamins throughout the night, so stock up!

2. WHILE YOU'RE DRINKING

Have another glass of water!

And do yourself a favor by choosing lighter-colored
liquors. Dark drinks like red wine, bourbon,
brandy, and whiskey contain higher concentrations of
substances called congeners, which act as
extra-toxic chemicals that your body has to eliminate.

And even though beer doesn't fit in this category, the
age-old rhyme "Beer before liquor, never been
sicker" does apply. Carbonation, like that in beer or
soda pop, increases the rate of absorption of
alcohol, getting you drunk quicker and causing a
worse hangover later.

3. AFTER THE BAR

MORE WATER. You'll be peeing all night, but it's better
than your head exploding tomorrow morning.

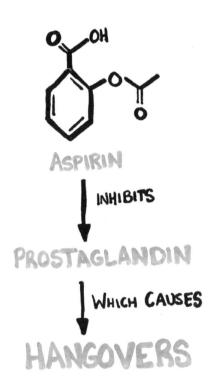

ASPIRIN

INHIBITS

PROSTAGLANDIN

WHICH CAUSES

HANGOVERS

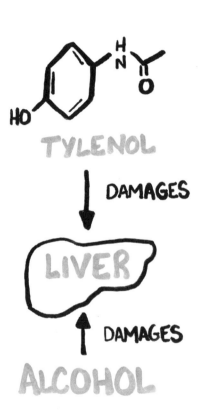

TYLENOL

DAMAGES

LIVER

DAMAGES

ALCOHOL

Aspirin before bed is also helpful;
it's noncaffeinated and inhibits
the release of prostaglandins,
which have been shown
to contribute to hangovers.

But stay away from acetaminophen
or paracetamol, aka Tylenol.
Paracetamol is extremely harsh on
your liver, and in combination with
alcohol can do some severe damage.

4. THE DREADED MORNING AFTER

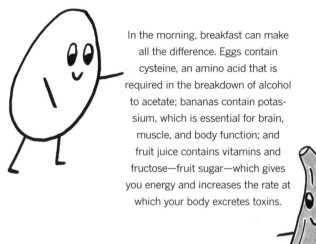

In the morning, breakfast can make all the difference. Eggs contain cysteine, an amino acid that is required in the breakdown of alcohol to acetate; bananas contain potassium, which is essential for brain, muscle, and body function; and fruit juice contains vitamins and fructose—fruit sugar—which gives you energy and increases the rate at which your body excretes toxins.

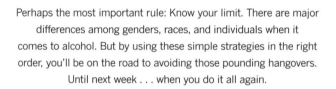

Perhaps the most important rule: Know your limit. There are major differences among genders, races, and individuals when it comes to alcohol. But by using these simple strategies in the right order, you'll be on the road to avoiding those pounding hangovers. Until next week . . . when you do it all again.

DREAMING, WAKING, NAPPING, SLEEPING

THE SCIENTIFIC POWER OF NAPS

Feeling groggy, tired, or unmotivated in the afternoon? Using coffee, soda, and energy drinks, we often try to push through the dreaded long day, yawning through the hours and fighting the fatigue. But it just so happens that the solution is the very thing we've been trying to avoid all along: sleep. Truth is, the power nap is perhaps the most effective way to rejuvenate your brain.

There are four main stages of your sleep cycle. The first two are relatively light sleep, while the third brings you into deep slumber. The final stage, known as rapid eye movement or REM, is where most of your dreams begin. The benefits of napping are tied to the length of time you're asleep.

10-30 MIN !

Naps ten to thirty minutes in length generally allow time only to enter the first stages. In stage 1, slow eye movement begins, and if awakened, you often feel as though you "didn't even sleep!" But as you continue into stage 2, your brain begins ignoring external stimuli that it deems nondangerous in order to relax you and give you a tranquil sleep. It also begins memory consolidation, in which information you learned is processed. Waking out of stage 2 sleep has shown benefits including increased productivity, higher cognitive functioning, enhanced memory, boosted creativity, and most important, feeling less tired.

30+ MIN

Beyond thirty minutes, you enter stage 3 and experience "sleep inertia" when awakened. Because your body is coming out of a deep sleep, your motor dexterity is decreased, while grogginess and the longing to go back to sleep increase. Many people falsely deem naps nonbeneficial for themselves, but the truth is they've simply napped too long.

As the benefits of napping are becoming clearer, many nap salons have opened throughout Japan where workers can pay to have a brief lunchtime nap on a daybed to increase alertness at work. So maybe it's about time we all start sleeping on the job a little more. Just tell your boss, "Science said so!"

THE SCIENCE OF MORNING WOOD

It affects grown men, little boys, and even males still in utero. You know, the ability to "pitch a tent" in the morning without any actual camping skills? Now that we've got the one obligatory joke out of the way, we can get to the hard facts of nocturnal penile tumescence, otherwise known as morning wood.

Morning erections are part of the normal sleep cycle and occur multiple times throughout the night.

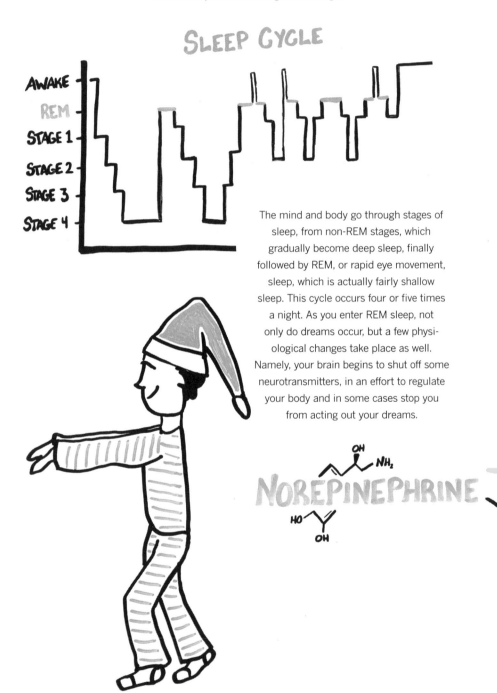

SLEEP CYCLE

AWAKE
REM
STAGE 1
STAGE 2
STAGE 3
STAGE 4

The mind and body go through stages of sleep, from non-REM stages, which gradually become deep sleep, finally followed by REM, or rapid eye movement, sleep, which is actually fairly shallow sleep. This cycle occurs four or five times a night. As you enter REM sleep, not only do dreams occur, but a few physiological changes take place as well. Namely, your brain begins to shut off some neurotransmitters, in an effort to regulate your body and in some cases stop you from acting out your dreams.

NOREPINEPHRINE

One of these neurotransmitters, norepinephrine, happens to be involved in the control of erections. Specifically, it causes vasoconstriction of the penis's blood vessels, actively preventing an erection—it's like a stop sign to blood flow.

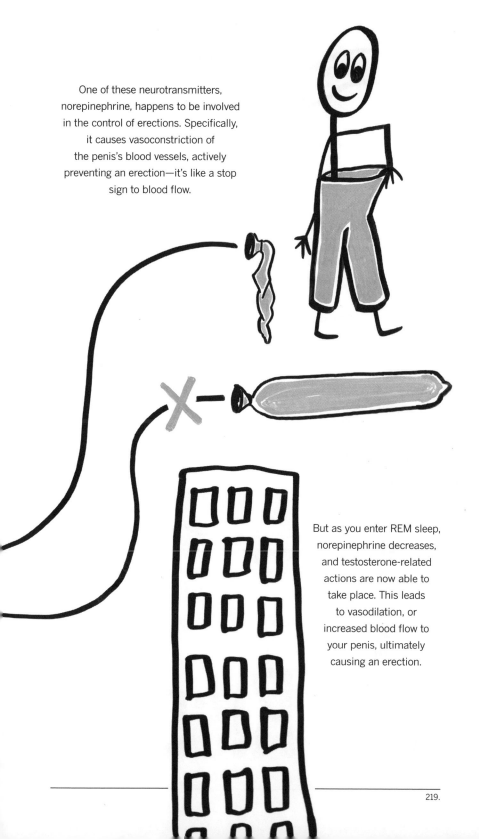

But as you enter REM sleep, norepinephrine decreases, and testosterone-related actions are now able to take place. This leads to vasodilation, or increased blood flow to your penis, ultimately causing an erection.

Why is this important? Well, similar to muscle and other tissue, at night, this extra blood increases oxygenation, serving as a system of repair and helping to maintain functionality.

But *why* do we often wake up to this pleasant surprise? Well, we often wake up just out of REM sleep and, as a result, see the remnants of this sleep stage.

RISE &

There is also evidence that a full bladder can contribute to morning wood. The increased bladder size through the night stimulates a region of the spinal cord, which can cause a "reflex erection."

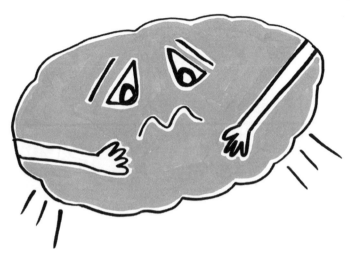

The physiological benefit of this is to prevent you from urinating in your sleep. But most men can attest to the . . . uh . . . difficulty proposed by this conundrum after waking in the morning.

SHINE

THE SCIENCE OF LUCID DREAMING

Dreams provide an escape from reality, into the mind, a realm where we seemingly have little control over what happens. But what if you could know you were dreaming and subsequently control the dream? It turns out, lucid dreaming is entirely possible, and with a little effort and practice even you can do it.

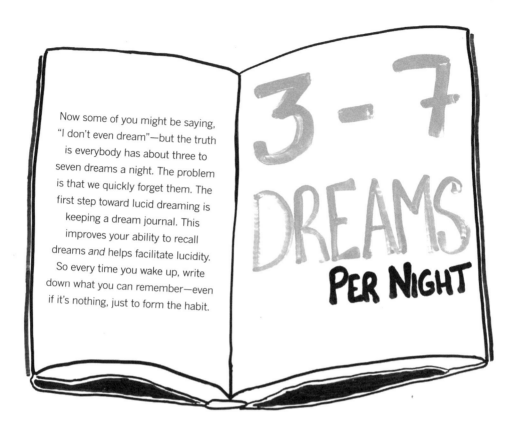

Now some of you might be saying, "I don't even dream"—but the truth is everybody has about three to seven dreams a night. The problem is that we quickly forget them. The first step toward lucid dreaming is keeping a dream journal. This improves your ability to recall dreams *and* helps facilitate lucidity. So every time you wake up, write down what you can remember—even if it's nothing, just to form the habit.

3 - 7 DREAMS PER NIGHT

The next step is performing reality checks. In a dream, something as simple as reading a sentence, counting your fingers, or checking the time can often go astray. Try it right now—look at the time, look away, and then look back. Assuming you aren't currently dreaming, the time probably stayed the same. However, in a dream, the time or the words you were reading often completely change. The key is to do these reality checks often, when you're awake. This way they become second nature, and when you're dreaming you're likely to perform the same test and realize that something is wrong.

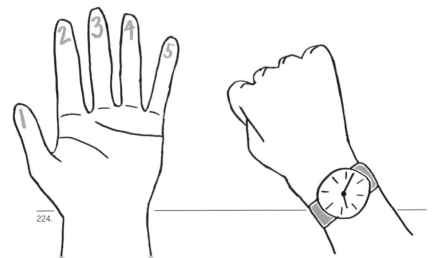

MILD TECHNIQUE

After this comes a technique known as mnemonically induced lucid dreams, or MILD. As you're falling asleep, begin to think of a recent dream and imagine yourself becoming lucid. The idea is to reinforce the intention to realize you're dreaming . . . in your dream. Keep repeating the phrase: "I will have a lucid dream tonight." The highest rates of success tend to come if you wake up in the middle of the night, get up for thirty minutes, and then go back to sleep with these intentions in mind.

WILD TECHNIQUE

Finally, once you have had success with MILD, you can attempt the advanced technique known as wake-induced lucid dreams, or WILD. The idea here is to keep your mind aware while your body falls asleep. Keep your body completely relaxed and don't move. The risk here is that you will experience sleep paralysis—a completely normal phenomenon that prevents your body from moving during your sleep. Except you'll be awake—which can be somewhat frightening.

The extra caveat with WILD is that during sleep paralysis the brain can play tricks on you, inducing strong feelings of fear and causing hallucinations of dark and scary figures approaching you. But don't worry—these are no more real than a bad dream!

226.

Scientific research into lucid dreaming has provided an insight into the location of metaconsciousness in the brain, provided opportunities for dream therapy and prevention of nightmares, and even begged the question if sleep and wakefulness are distinct events or part of a continuum.

After all, dreaming of doing something is almost equivalent to actually doing it, when looking at the functional system of neuronal activity in your brain. So are you *sure* you're not dreaming?

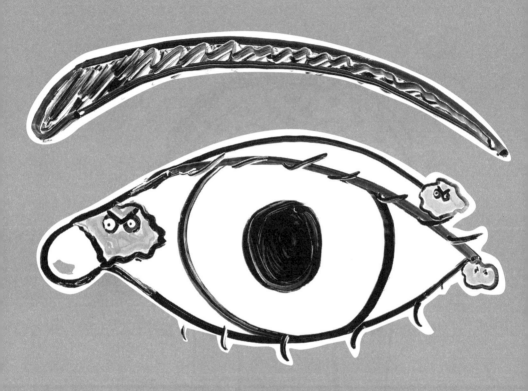

WHAT ARE EYE BOOGERS?

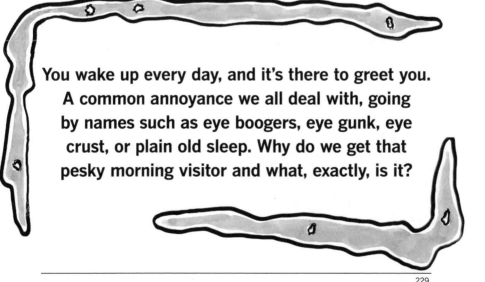

You wake up every day, and it's there to greet you. A common annoyance we all deal with, going by names such as eye boogers, eye gunk, eye crust, or plain old sleep. Why do we get that pesky morning visitor and what, exactly, is it?

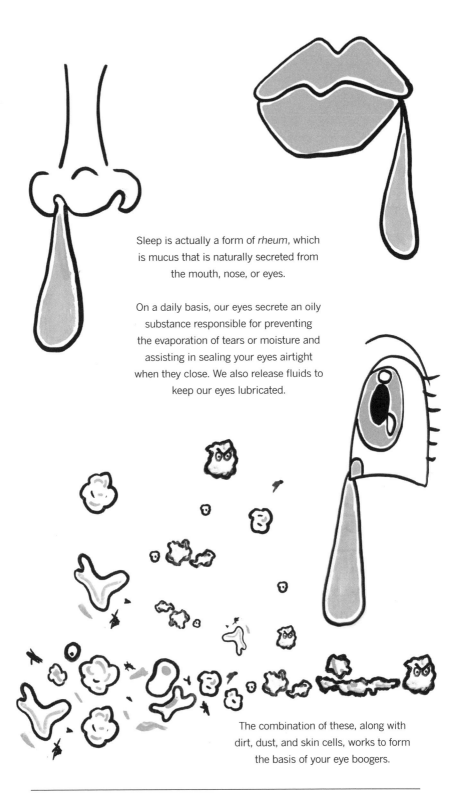

Sleep is actually a form of *rheum*, which is mucus that is naturally secreted from the mouth, nose, or eyes.

On a daily basis, our eyes secrete an oily substance responsible for preventing the evaporation of tears or moisture and assisting in sealing your eyes airtight when they close. We also release fluids to keep our eyes lubricated.

The combination of these, along with dirt, dust, and skin cells, works to form the basis of your eye boogers.

When you're awake, the rheum is flushed away by your blinking eyelids. But when you're asleep, you don't blink, so rheum gathers in the inner corners and lash lines of your eyes, morphing overnight into what we call *sleep*. The eye rheum is "cooked" with a combination of the dry air around you and your own body heat.

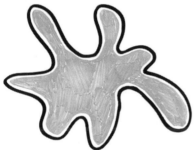

Depending on the amount of liquid in the rheum, your body temperature, and the time the liquid has to evaporate while sleeping, those little eye nuggets will be either crusty and hard or wet and gooey by the time you wake.

Waking up with a collection of sleep in your eyes is a pretty ordinary occurrence, but there are some instances when our eyes tend to get exceedingly cruddy. When you catch a cold, you are prone to excessive and watery eye gunk.

If you are experiencing a lot of nasal congestion, the close connection between your eyes and nose allows mucus in your nasal passages to flow back into your eyes while sleeping, and consequently, an above-average amount of mucus results in heavier deposits of sleep.

In the inner corner of your eye, you will find a tiny, pink, crescent-moon shaped structure called the "plica semilunaris."

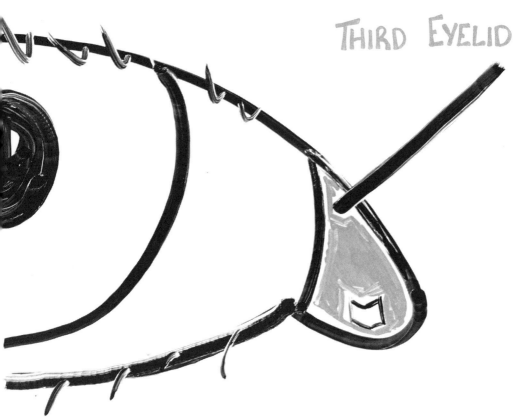

THIRD EYELID

This pink part of our eye is rich in cells that create mucus, an important contribution to the formation of sleep! It could also be the remnant of a third eyelid that our evolutionary ancestors may have had, but in humans has lost its function over time.

A third eyelid can be found in some mammals
such as cats, seals, and polar bears.
This translucent membrane slides horizontally
across the animal's eye, for protection
under water as well as a defense against the
sun. It helps keep the eye clean, like a
windshield wiper blade, spreading tears across
the eye and eliminating debris.

Even though wiping sleep out of
your eyes may be a bothersome
part of your morning routine, those
little buggers play a crucial role
in good eye health. By guiding the
rheum into the corners of your eyes,
the corneas are protected from
potentially damaging debris! So,
finding a little crud after waking from
a slumber is just a sign that your
eye is trying to keep you healthy.

SHOULD YOU USE THE SNOOZE BUTTON?

The snooze button: one of man's best inventions . . . until nine minutes later when the dreaded alarm strikes again. Except now you feel even more tired, so do you hit it again? Are those extra minutes really helping you at all, or is it the beginning of a never-ending cycle that ends in you being late . . . and still dead tired?

In an unimaginable world without alarm clocks, our bodies would simply wake up naturally. Seems crazy, right? But our bodies have many chemical mechanisms in place not only to put us to sleep but to wake us up as well.

The body begins preparing in the hour before you naturally wake up. Body temperature rises, sleep becomes lighter, and hormones such as dopamine and cortisol are released, which give you energy to start your day. But the problem with alarms is that they often interrupt your sleep cycle and cut these processes short, particularly if you don't have a regular sleep rhythm, or schedule. The alarm goes off, but your body isn't quite ready. This groggy and tired state is known as sleep inertia, and its strength is related to which sleep stage you're waking up out of; the deeper the sleep, the more potent the sleep inertia. And so the snoozing begins.

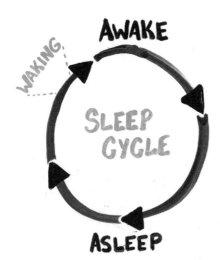

AWAKE

WAKING

SLEEP
CYCLE

ASLEEP

But the snooze button can do more damage than good. As you fall back asleep, your body may restart its sleep cycle and enter into *deeper* sleep stages.

ZZ z z z z ZZZZ zz

So instead of your body prepping to wake up, it's going in the opposite direction! As a result, the second alarm may cause you to feel even more tired. And so continues the vicious cycle. Ultimately, you're better off setting your alarm later and not interrupting your sleep.

X   10   10   10

VS

✓   30

Many studies have found that fragmented sleep is much less restorative and leads to sleepiness-related daytime impairment. So by breaking up those last thirty minutes or so of sleep, you're more likely to feel tired and perform poorly during the day.

What else can you do? Try adopting a more regular sleep schedule. Being tired is a product not only of sleep deprivation and waking up out of deep sleep but also of lacking a consistent schedule. Your body loves predictability.

Wake up at the same time every morning—including the weekends—and after a few weeks, your body should adapt to the timing and be less inclined to require an alarm in the first place.

And if you do wake up feeling a little tired, resist
the snooze temptation and just get up.
As the saying goes, "You snooze, you lose."

YOU SNOOZE, YOU LOSE!

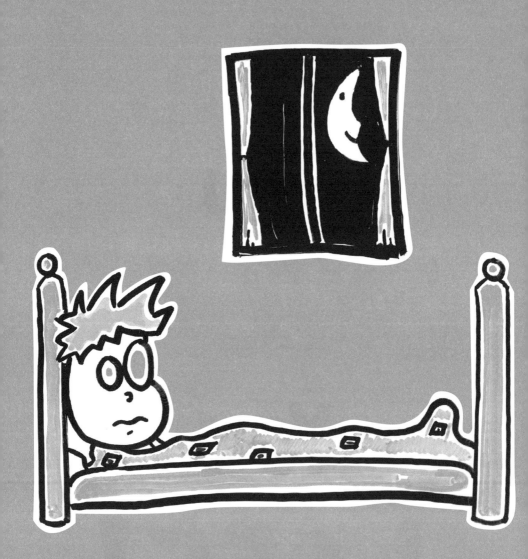

WHAT IF YOU
STOPPED
SLEEPING?

Ahhh, sleep! You can never have enough of it, it seems. In fact, sometimes it literally feels like you aren't getting enough. But what if you stopped sleeping altogether?

Strangely, science understands relatively little about why we sleep or how it evolved in the first place. After all, lying unconscious for hours on end while predators lurk hardly seems advantageous or smart.

But we have discovered a few correlations. For example, adults who sleep six to eight hours a night tend to live longer! Excessive sleep, however, can lead to medical problems including cardiovascular disease and diabetes. Similarly, chronic sleep deprivation has been linked to aspects of cardiovascular disease, obesity, depression, and even brain damage.

But what would happen if you stopped sleeping *right now*? Well, after your first sleepless night your mesolimbic system becomes stimulated and dopamine runs rampant. And this may actually trigger some extra energy, motivation, positivity, and even sex drive.

Sounds appealing, but it's a slippery slope! Your brain slowly begins to shut off the regions responsible for planning and evaluating decisions, leading to more impulsive behavior.

Once exhaustion sets in, you'll find yourself with slower reaction times and reduced perceptual and cognitive functions. After a day or two of no sleep, your body loses its ability to properly metabolize glucose, and the immune system stops working as well.

In some cases, three days of no sleep has led to hallucinations.

Care about how you look? Studies have shown a direct correlation between sleep deprivation and a person's perceived beauty. That is to say, sleep-deprived individuals appeared less healthy *and* less attractive than when they were well rested.

The longest scientifically documented case of being awake was 264 hours, or 11 days. And while this man did develop problems with concentration, perception, and irritability, the surprising truth is that he suffered no serious long-term health effects.

NO LONG TERM HEALTH EFFECTS

In fact, no individuals under these documented conditions experienced medical, physiological, neurological, or psychiatric problems. But these are limited studies, and this doesn't mean permanent damage couldn't be inflicted with more time.

Sleep deprivation experiments on rats, for example, generally led to death after about two weeks. But scientists aren't totally sure if they died from lack of sleep or from the stress of constantly being woken up.

Perhaps we should look at fatal familial insomnia for an answer. This is a rare genetic disease of the brain that causes progressively worsening insomnia or sleeplessness, leading to hallucinations, dementia, and, ultimately, death.

100 PEOPLE

This disease has been documented in only around one hundred people in the world, but their average survival span was around eighteen months after symptoms appeared. Over time, the lack of sleep becomes worse, and the body's organs begin to shut down.

So while lack of sleep won't necessarily kill you quickly, continual sleep deprivation will have a negative effect on your body.

SLEEP TIGHT
... BUT NOT TOO MUCH!

Acknowledgments

First and foremost, this book could not have been created without our friend
and colleague Jess Carroll and her creative direction, brilliant illustrations,
and continued support with all aspects of AsapSCIENCE. Without her, this
book would have taken even longer, and used a heck of a lot more
stick figures. Jess—you are the atomic glue that is holding us together!
And of course we are grateful to our designer Brian Chojnowski, who
took those drawings and our ramblings and beautified them all into such a cool
book! Thanks to our amazing research team, Jess Gemin and
Gillian Brown, who helped with the essential, grueling process of reading
scientific journals, brainstorming ideas, and fact checking.

We cannot forget the girl crazy enough to start this whole process—our
literary agent, Sasha Raskin, who helped us conceive the book and
really inspired the idea in us, which of course lead us to the best, most lovely
editor we could have asked for. Shannon Welch: your passion, guidance,
and (most important) patience with us was heaven-sent. Seriously, we
couldn't be more grateful for you in this whole process—you kept us sane,
but also in line!

Thanks to all of our friends who put up with our science obsession—most
of all our roommates, Brian and Sarah, who hear it on a daily basis.
And of course, thank you to our families: Anne, Bob, and
Gill Brown and Wendy, Phil, Kim, Matt, and Mike Moffit.
You've all been eternally supportive of and inspiring to us,
and we would not be the people we are without you.

For more information about
AsapSCIENCE and the sources for all
the science in this book, please visit
youtube.com/AsapSCIENCE.